I0126793

# Journal of
# South Carolina Water Resources

Volume 5, Issue 1      2018

## Contents

# Foreword

## Dawn Anticole White, M.M.C.

*Journal of South Carolina Water Resources* Managing Staff Editor

Oftentimes considered fodder for small talk, the weather has become an increasingly complex discussion due to changing climatic effects. With significant episodes of drought and high-intensity rainfall occurring more frequently, weather and climate are necessary conversation topics that are shaping new approaches to water resources management. Melissa Griffith, assistant state climatologist for the South Carolina State Climatology Office (SCO), summarized recent weather events that spanned both ends of the hydrological spectrum:

Over the last 4 years, South Carolina has had its share of water-related weather and climate issues. On the heels of the disastrous flooding of October 2015, Hurricane Matthew produced up to 20 in. of rainfall in the Waccamaw River Basin a year later, causing record flooding in portions of the coastal plain. At the same time, the Upstate was dealing with severe drought conditions that fueled the Pinnacle Mountain fire in November 2016, the most destructive fire on record for the region. In 2017, incipient and moderate drought conditions continued to affect the state, and rainfall from Hurricane Irma in September helped ease some of those issues, without the significant flooding. Despite starting 2018 with 8 in. of snow along the coast, the drought conditions persisted until above-normal rainfall in May led to the first "drought-free" declaration for the entire state since July 2016. Unfortunately, drought conditions started to emerge again in the summer of 2018, and South Carolina was dealt yet another blow in September—the third major flood in 4 years—as Hurricane Florence dropped up to 2 ft. of rain across the Pee Dee Watershed. The excessive rainfall produced prolonged flooding on rivers that broke previous records set by Hurricane Matthew just 2 years prior.

In response to these extreme events, new initiatives are emerging. From late 2017 into early 2018, the SCO, along with the Carolinas Integrated Sciences and Assessments (CISA) program and Clemson University's South Carolina Water Resources Center (SCWRC), conducted the Climate Connections Workshop Series (http://www.dnr.sc.gov/ccworkshops/). The workshops, entitled "Weathering the Storm: Impacts of Extremes on South Carolina's Natural and Built Environment," provided information about Hurricanes Matthew (2016) and Irma (2017), as well as the 2016 drought and wildfires, state and local responses, and how lessons learned are being incorporated into future planning. In addition, based on participant feedback from the 2017 South Carolina Drought and Water Shortage Tabletop Exercise, the SCO and CISA launched a new drought website (http://www.scdrought.com/). The portal fulfills the need for improved education and awareness with information about drought monitoring, response, and planning. At the end of 2017, the first Groundwater Availability Assessment meetings were coordinated by the SCWRC in the inner and outer coastal plains, and the center is also partnering with the South Carolina Department of Natural Resources and the U.S. Army Corps of Engineers to develop methods to project future water demand for the state. The methodology process began in August 2018 and will be completed in 2019 with draft projections applied to the Savannah River Basin as a pilot. Additionally, to assist with gathering water usage data, Clemson University Public Service and Agriculture (PSA) and Clemson Cooperative Extension launched the South Carolina Agricultural Water Use and Irrigation Survey in 2017 to quantify agricultural water use and irrigation practices. Another notable development over the past year was the formation of the State Water Planning Process Advisory Committee, whose members are devising a framework to guide river basin councils with the development of regional water plans, which will be incorporated into the new State Water Plan.

Many were able to share and discuss information about these initiatives and more at the 2018 South Carolina Water Resources Conference, "Ten Years of Water Science and Policy: Engaging in South Carolina's Water Future" held in October. Hosted by Clemson PSA and coordinated by the SCWRC, the conference again brought together over 300 attendees for a wide range of water science, policy, and application presentations. The engaging plenary speakers included legislators and representatives from federal and state management and regulatory agencies, as well as South Carolina Public Radio NatureNotes host, Rudy Mancke.

There has been continued progress to improve water management and planning and make weather and climate discussion a priority. The 6 articles in this issue cover a variety of topics including evapotranspiration, toxicity, groundwater, stakeholder communications, and water use. Wider understanding of these complex issues is critical to protecting water resources—however prolonged the conversations are. We sincerely thank the authors for their contributions and the reviewers for their expertise.

Journal of South Carolina Water Resources, Volume 5, Issue 1, Pages 3–24, 2018

# Assessment of Spatial and Temporal Variation of Potential Evapotranspiration Estimated by Four Methods for South Carolina

Devendra M. Amatya[1], Augustine Muwamba[2], Sudhanshu Panda[3], Timothy Callahan[2], Scott Harder[4], and C. Alex Pellett[4]

AUTHORS: [1]USDA Forest Service, 3734 Highway 402, Cordesville, SC 29434. [2]Department of Geology and Environmental Sciences and Mathematics, College of Charleston, 66 George Street, Charleston, SC 29424. [3]College of Engineering, University of North Georgia, 82 College Cir, Dahlonega, GA 30597. [4]South Carolina Department of Natural Resources, 311 Natural Resources Drive, Clemson, SC 29631.

**Abstract.** Given South Carolina's ongoing water planning efforts, in this study, we evaluated seasonal and annual potential evapotranspiration (PET) using measured Class A pan evaporation (PE) and 3 widely used estimation methods for the state with 3 distinct physiographic regions (Coastal, Piedmont, and Mountain). The methods were temperature-based Hargreaves-Samani (H-S), radiation-based Priestley-Taylor (P-T), and process-based Penman-Monteith (P-M). The objectives of the study were to (a) describe seasonal and temporal distribution of PET by all methods, (b) quantify differences among PET methods, and (c) identify relationships between monthly PE and estimated PET by each method. Daily weather variables from 59 National Oceanic and Atmospheric Administration weather stations distributed in the 3 regions of South Carolina (SC) were used to estimate daily PET for an 18-year period (1998–2015). Net radiation was estimated using modeled solar radiation values for weather stations. The average annual H-S PET values adjusted with the empirical radiation factor (KT) and the average annual P-T PET values for 1998–2015 were 1,232 ± 9, 1,202 ± 11, and 1,115 ± 10 mm and 1,179 ± 10, 1,137 ± 11, and 1,082 ± 11 mm, respectively, for the Coastal, Piedmont, and Mountain regions. Both the mean annual H-S and P-T PET for the Mountain region were significantly ($\alpha = 0.05$) lower than for the Coastal and Piedmont regions. The mean annual P-T PET for the Coastal region was significantly ($\alpha = 0.05$) greater than that for the Piedmont. Regional differences showed that estimated PET for 1998-2015 was greatest in the Coastal and lowest in the Mountain region. Comparison of all 3 methods using only common 8-year data showed mean annual P-M PET, varying from 1,142 mm in the Piedmont to 1,270 mm in the Coastal region, was significantly higher than both the H-S and P-T PET in both regions. The greatest mean monthly H-S and P-T PET values were observed in June and July. Statistical evaluation using Nash–Sutcliffe efficiency and percent bias showed a slightly better agreement of H-S PET with both the measured PE as well as the P-M method, followed by the P-T. However, the P-T method yielded a close to unity slope and slightly higher $R^2$ than the H-S PET when compared with the PE. The P-T PET method that uses both the temperature and radiation data may be preferred for SC with a humid climate dominated by forest land use, given more rigorous ground-truthing of modeled solar radiation as data become available. Surface interpolation algorithm, inverse distance weighted, was used to spatially map both the distributed H-S and P-T PET for the state. Results from this study can be used to support several components of the ongoing water planning efforts in SC.

## INTRODUCTION

Evapotranspiration (ET) is a major component of the water cycle and influences runoff, soil water storage, groundwater recharge, biodiversity, and the global climate system (McMahon et al., 2013; Tegos et al., 2015; Sun et al., 2011). Potential evapotranspiration (PET) is a common measure of evaporative demand defined as the rate of evaporation that would occur from soil and plant surfaces with unlimited water supply and no resistance to transfer of water (Hember et al., 2017). Numerous methods and models for estimating PET have been developed that range from pan evaporation (PE) to the parameter-intensive, physically based Penman-Monteith (P-M; Monteith, 1965) method to temperature-only-based methods to many other methods of varying complexity (McMahon et al., 2013).

Some of the widely-used temperature-based PET methods, which rely on temperature as the primary climate variable to estimate PET directly or indirectly, include Thornthwaite (1948), Hamon (1963), and modified Hargreaves–Samani (H-S; Hargreaves & Samani, 1985). Because PET is also controlled by other climatic variables like solar radiation, humidity, and wind speed, and owing to advances in computing technology, there has been a tremendous effort in the last few decades to develop process-based PET models (Allen et al., 1998; Marek et al., 2016). Furthermore, the interaction of vegetative parameters like leaf area index and stomatal conductance with the microclimate including aerodynamic resistance has also been shown to be important for addressing the PET for a given surface resistance in the process-based PET models (Brauman et al., 2012; McKinney & Rosenberg, 1993). In recent years, the U.N. Food and Agricultural Organization (FAO)-56 Penman–Monteith model (Allen et al., 1998), which is a slight modification of the original P-M method, has been used globally as a reference ET (REF-ET or $ET_0$) for a standard grass from which to compare crop ET of all other crops under nonstressed conditions (Allen et al., 1998; Allen et al., 2006; Amatya & Harrison, 2016; Amatya et al., 1995; Cai et al., 2007; FAO, 1990; Lima et al., 2013; Lopez-Moreno et al., 2009; McMahon et al., 2013; Rao et al., 2011; Wang et al., 2015; Raziei & Pereira., 2013). Shevenell (1996) used measured temperature and the calculated ratio of total to vertical radiation to estimate monthly PET at 125 weather stations in Nevada, most of which are near valley floors at elevations ranging from 393 to 2,287 m. The author reported that the calculated values were found to be well correlated ($R^2 = 0.91–0.99$, slopes near 1.0) with monthly PE measurements at 8 sites in Nevada.

There have been only limited studies conducted to assess the short- or long-term PET for South Carolina. Barker and Pernik (1994) noted that although regional maps of actual ET were not available, 2 published maps of PET were available that covered the Southeastern Coastal Plain aquifer system including South Carolina. The first one, by Geraghty et al. (1973), provided a PET map used by the U.S. Department of Agriculture (USDA) Forest Service to estimate ET from the oak–hickory–pine forests in the southeastern United States. Their estimates indicated that PET ranges from 36 in. (900 mm) per year along the northern edge to 40 in. (1,000 mm) along the southern edge. The second one is Hamon's (1963) map of PET based on air temperature, saturation vapor pressure, and daytime hours for the eastern United States. Barker and Pernik (1994) used the Hamon's PET method to estimate ET for their study areas in the Southeastern Coastal Plain aquifer system because they found that the Hamon's results were close to those estimated from the Thornthwaite (1948) and Penman (1948) methods. Young (1968) documented a 3-year estimate of Thornthwaite-based PET using temperature data at the USDA Forest Service's Santee Experimental Forest (SEF) headquarters in coastal South Carolina. Lu et al. (2005) contrasted 6 commonly used PET models and quantified the long-term PET across a physiographic gradient of 36 watersheds in the southern United States including 1 at the SEF. Three temperature-based (Hamon, 1963; H-S, Hargreaves & Samani, 1985; Thornthwaite, 1948) and 3 radiation-based (Makkink, 1957; P-T, Priestley & Taylor, 1972; Turc, 1961) PET methods were compared. The authors concluded that, in general, the P-T, Turc, and Hamon methods performed better than the other PET methods and that the ET values with temperature-based methods were more variable than those obtained from solar radiation-based methods. Later, Harder et al. (2007) compared 3 methods (Hamon, 1963; P-M, Monteith, 1965; Thornthwaite, 1948) to estimate PET using data from a standard weather station above a grass reference also at the SEF. The authors also found that the temperature-based Thornthwaite and Hamon methods yielded results close to those by the P-M method. Amatya and Harrison (2016) evaluated 5 different methods (H-S, Hargreaves & Samani, 1985; P-M, Monteith, 1965; P-T, Priestley & Taylor, 1972; Thornthwaite, 1948; Turc, 1961) to estimate daily and monthly PET for a pine and hardwood forest in coastal South Carolina and found P-T and H-S PET matching closely with the P-M PET. Similarly, the P-M based reference ET estimates using 9-year (2001–2009) data were recently reported for 21 stations in South Carolina for their potential use in agricultural water management by the Natural Resources Conservation Service (NRCS, 2016).

All of these studies were more site specific and with limited data. Furthermore, most of the aforementioned PET studies have been conducted for a well-watered standard grass reference that may not well represent evaporation from large, open-water bodies like reservoirs and lakes often used for multipurpose water management (Rosenberry et al., 2007). A pan coefficient is generally applied to measurements of PE from a National Weather Service Class A pan to account for heat transfer through the sides and bottom of the pan for estimating open water evaporation (OWE; Hember et al., 2017) of large water bodies with deep storage (Jensen et al., 1990). There are some empirical methods in the literature to derive pan coefficients using some measured climatic variables like wind speed, relative humidity, and upwind fetch distance (Grismer et al., 2002; Irmak et al., 2002). Phillips et al. (2014) compared remote sensing estimates of lake evaporation with PE measurements along the Savannah River Basin and attributed seasonal variabilities in lake evaporation to seasonal variations in temperatures. Recently, CDM Smith (2016) reported a long-term assessment of H-S-based PET compared with pan and OWE for some limited locations in South Carolina.

As the issues of water supply, reservoir water management, drought, irrigation and crop water use, and land use change are becoming of a societal concern given the pressures of urbanization and climate change (Lackstrom et al., 2016; Mizzell et al., 2014; Roehl & Conrads, 2015), there is a growing need for more reliable operational methods and tools to assess long-term ET and PET to support sound management decisions on water resources. Therefore, our objectives in this study were to (a) describe spatially distributed seasonal and annual PET by 3 widely used methods (H-S, originally developed for cool-season grass in subhumid to arid western United States; P-T, originally developed for rain-fed grassland in Australia and United States; and P-M, for a standard 12-cm-high grass at all locations), (b) quantify differences in computed PET among the 3 methods in each region and among 3 regions for each PET method, (c) compare each of the H-S and P-T PET methods with the standard grass reference-based P-M PET for all sites, and (d) examine the relationships between monthly PE and PET by each of the 3 methods for the state of South Carolina. Although the H-S and P-T PET methods were originally developed for only 7- and 10-day

periods, respectively (Jensen et al., 1990), calculations on a daily time step were performed to obtain the monthly values for all the 3 methods in this study. Spatially interpolated GIS maps were developed using both the H-S and P-T methods.

## MATERIALS AND METHODS

### STUDY SITES

The climatological (weather) stations used as study sites for the spatiotemporal assessment of PET are shown in Figure 1; their names and characteristics are listed in Table 1.

A total of 59 stations distributed across the physiographic areas (Coastal, 31; Piedmont, 24; Mountain, 4) are maintained by National Oceanic and Atmospheric Administration (NOAA). The ranges of elevation (above mean sea level) were approximately 2.4–137.2 m, 70.1–319.4 m, and 298.7–975.4 m, for stations in the Coastal, Piedmont, and Mountain regions, respectively, whereas

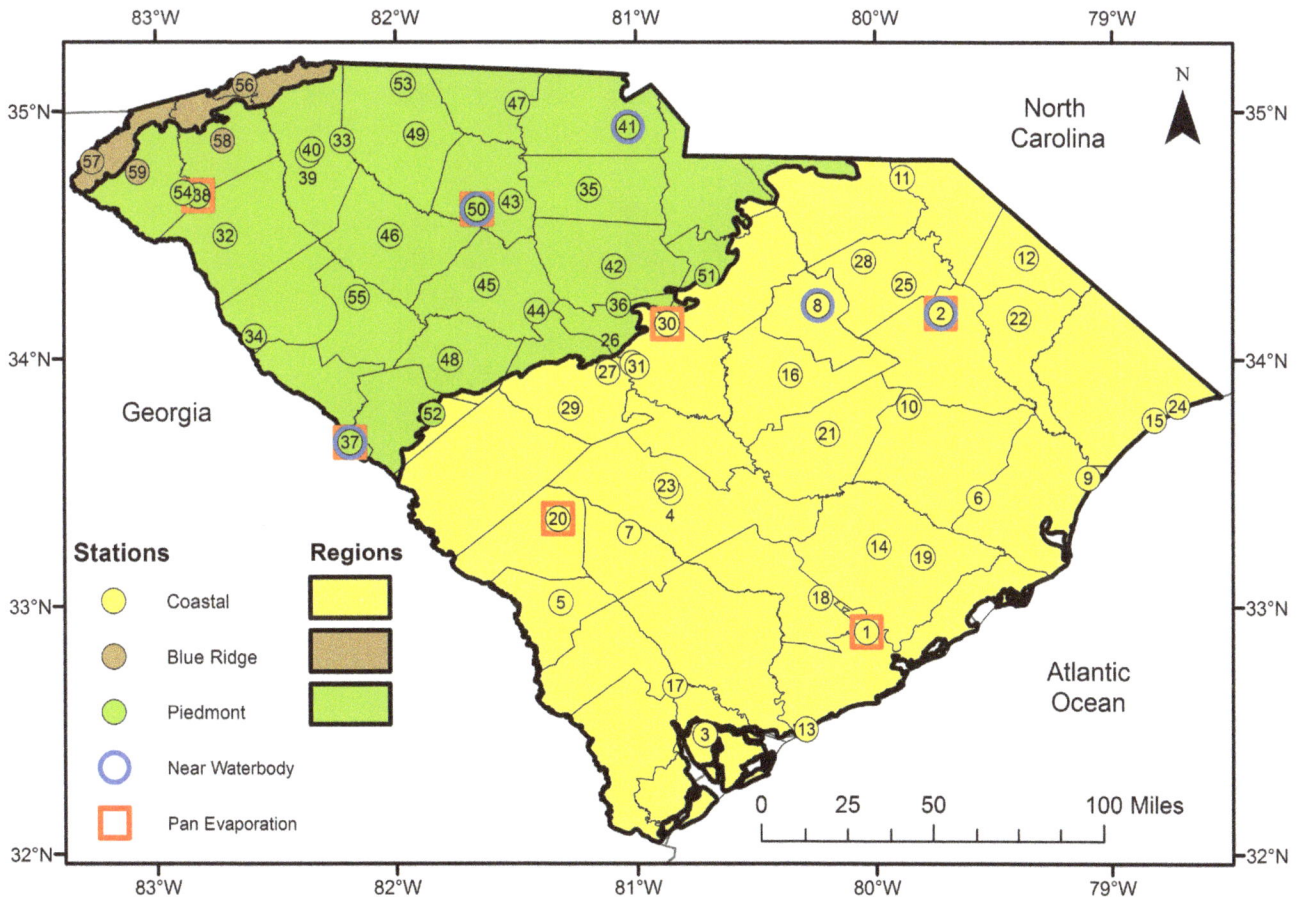

**Figure 1.** Spatial distribution of 59 National Oceanic and Atmospheric Administration weather stations selected for this study. Circled stations are near open water bodies (lakes and reservoirs).

Sources:

Menne, Matthew J., Imke Durre, Bryant Korzeniewski, Shelley McNeal, Kristy Thomas, Xungang Yin, Steven Anthony, Ron Ray, Russell S. Vose, Byron E.Gleason, and Tamara G. Houston (2012): Global Historical Climatology Network - Daily (GHCN-Daily), Version 3.21. NOAA National Climatic Data Center. doi:10.7289/V5D21VHZ [6/25/2015].

Omernik, J.M. and G.E. Griffith. 2014. Ecoregions of the conterminous United States: evolution of a hierarchical spatial framework. Environmental Management 54(6):1249-1266.

**Table 1.** NOAA weather stations and their characteristics in Coastal (C), Piedmont (P), and Mountain (M) regions.

| Station | City | Lat, °N | Long, °W | Elev, m | Station ID (US) | Reg | H-S PET, mm | P-T PET, mm |
|---|---|---|---|---|---|---|---|---|
| 1. Charleston AP[a] | Charleston | 32.90 | −80.04 | 12.2 | W00013880 | C | 1,206 ± 6[b] | 1,205 ± 11[c] |
| 2. Florence AP[a] | Florence | 34.19 | −79.72 | 44.5 | W00013744 | C | 1,210 ± 9 | 1,168 ± 11 |
| 3. Beaufort MCAS[d] | Beaufort | 32.48 | −80.72 | 11.3 | W00093831 | C | 1,199 ± 9 | 1,235 ± 11 |
| 4. Orangeburg MU AP | Orangeburg | 33.46 | −80.86 | 60 | W00053854 | C | 1,251 ± 17 | 1,203 ± 11 |
| 5. Allendale 2 NW | Allendale | 33.02 | −81.32 | 54.9 | C00380126 | C | 1,305 ± 11 | 1,231 ± 11 |
| 6. Andrews | Andrews | 33.44 | −79.57 | 10.7 | C00380184 | C | 1,224 ± 10 | 1,164 ± 11 |
| 7. Bamberg | Bamberg | 33.30 | −81.03 | 50.3 | C00380448 | C | 1,241 ± 14 | 1,187 ± 12 |
| 8. Bishopville 1ENE | Bishopville | 34.22 | −80.24 | 75.9 | C00380736 | C | 1,216 ± 10 | 1,158 ± 11 |
| 9. Brookgreen Gardens | Murrells Inlet | 33.52 | −79.10 | 6.1 | C00381093 | C | 1,109 ± 9 | 1,194 ± 11 |
| 10. Cades 4W | Lake City | 33.81 | −79.86 | 24.4 | C00381241 | C | 1,291 ± 10 | 1,183 ± 10 |
| 11. Dillon | Dillon | 34.41 | −79.36 | 35.1 | C00382386 | C | 1,224 ± 16 | 1,159 ± 11 |
| 12. Edisto BE ST PA | Edisto Beach | 32.51 | −80.29 | 2.4 | C00382730 | C | 1,067 ± 11 | 1,232 ± 10 |
| 13. Moncks Corner 4N | Moncks Corner | 33.24 | −79.99 | 14.9 | C00385946 | C | 1,242 ± 9 | 1,216 ± 11 |
| 14. Myrtle Beach[d] | Myrtle Beach | 33.75 | −78.82 | 11.9 | C00386153 | C | 1,089 ± 10 | 1,171 ± 11 |
| 15. Sumter | Sumter | 33.94 | −80.36 | 53.9 | C00388440 | C | 1,199 ± 13 | 1,151 ± 11 |
| 16. Yemassee | Yemassee | 32.68 | −80.84 | 13.4 | C00389469 | C | 1,409 ± 8 | 1,183 ± 13 |
| 17. Summerville 4W | Summerville | 33.04 | −80.23 | 19.8 | C00388426 | C | 1,257 ± 12 | 1,180 ± 10 |
| 18. Santee[f] | Cordesville | 33.20 | −79.80 | 14 | | C | 1,390 ± 21 | 1,132 ± 17 |
| 19. Blackville 3W[e] | Blackville | 33.36 | −81.33 | 96.6 | W00063826 | C | 1,252 ± 21 | 1,182 ± 11 |
| 20. Manning | Manning | 33.70 | −80.20 | 30.5 | C00385493 | C | 1,272 ± 13 | 1,186 ± 11 |
| 21. Marion | Marion | 34.17 | −79.39 | 22.9 | C00385509 | C | 1,214 ± 13 | 1,165 ± 10 |
| 22. Orangeburg | Orangeburg | 33.49 | −80.87 | 54.9 | C00386527 | C | 1,303 ± 13 | 1,201 ± 11 |
| 23. N. Myrtle Beach AP[d] | Myrtle Beach | 33.81 | −78.72 | 9.8 | W00093718 | C | 1,062 ± 11 | 1,157 ± 11 |
| 24. Darlington[d] | Darlington | 34.30 | −79.88 | 45.7 | C00382260 | C | 1,245 ± 10 | 1,174 ± 11 |
| 25. Columbia Met. AP[d] | Columbia | 33.95 | −81.12 | 68.6 | W00013883 | C | 1,230 ± 10 | 1,171 ± 12 |
| 26. Hartsville | Hartsville | 34.40 | −80.05 | 56.4 | C00383990 | C | 1,233 ± 11 | 1,137 ± 11 |
| 27. Pelion 4N | Pelion | 33.80 | −81.27 | 137.2 | C00386775 | C | 1,244 ± 10 | 1,167 ± 11 |
| 28. Sandhill Research[a] | Elgin | 34.14 | −80.87 | 134.1 | C00387666 | C | 1,234 ± 9 | 1,166 ± 11 |
| 29. Columbia Owens AP[d] | Columbia | 33.97 | −81.00 | 64.6 | W00053867 | C | 1,223 ± 11 | 1,180 ± 12 |
| 30. Cheraw | Cheraw | 34.73 | −79.88 | 42.7 | C00381588 | C | 1,181 ± 9 | 1,117 ± 11 |
| 31. Columbia University | Columbia | 33.98 | −81.02 | 73.8 | C00381944 | C | 1,324 ± 10 | 1,203 ± 12 |
| **Mean** | | | | | | **C** | **1,232 ± 9** | **1,179 ± 10\*** |
| 32. Cedar Creek 2E | Blythewood | 34.22 | −81.07 | 103 | C00381479 | P | 1,280 ± 14 | 1,121 ± 10 |
| 33. Anderson Co. AP | Anderson | 34.50 | −82.71 | 231.6 | W00093846 | P | 1,164 ± 11 | 1,154 ± 11[c] |
| 34. Greenville-Spartanburg AP[d] | Greer | 34.88 | −82.22 | 287.4 | W00003870 | P | 1,138 ± 11 | 1,133 ± 12 |
| 35. Calhoun Falls | Calhoun Falls | 34.09 | −82.59 | 161.5 | C00381277 | P | 1,210 ± 11 | 1,158 ± 11 |
| 36. Chester 1SE | Chester | 34.68 | −81.20 | 170.7 | C00381633 | P | 1,201 ± 10 | 1,109 ± 11 |
| 37. Clarks Hill 1W[e] | Clarks Hill | 33.66 | −82.19 | 115.8 | C00381726 | P | 1,280 ± 20 | 1,174 ± 11 |
| 38. Clemson Univ.[e] | Clemson | 34.66 | −82.82 | 251.2 | C00381770 | P | 1,172 ± 11 | 1,147 ± 11 |
| 39. Greenville | Greenville | 34.82 | −82.36 | 292.6 | C00383735 | P | 1,093 ± 9 | 1,115 ± 11 |
| 40. Greenville D AP[d] | Greenville | 34.85 | −82.35 | 319.4 | W00013886 | P | 1,122 ± 10 | 1,151 ± 11 |
| 41. Winthrop Univ. | Rock Hill | 34.94 | −81.03 | 210.3 | C00389350 | P | 1,162 ± 10 | 1,139 ± 11 |

| Station | City | Lat, °N | Long, °W | Elev, m | Station ID (US) | Reg | H-S PET, mm | P-T PET, mm |
|---|---|---|---|---|---|---|---|---|
| 42. Winnsboro | Winnsboro | 34.37 | −81.09 | 161.5 | C00389327 | P | 1,205 ± 12 | 1,130 ± 10 |
| 43. Santuck | | 34.64 | −81.52 | 158.1 | C00387722 | P | 1,206 ± 12 | 1,141 ± 12 |
| 44. Little Mountain | Little Mountain | 34.19 | −81.41 | 216.7 | C00385200 | P | 1,159 ± 11 | 1,150 ± 11 |
| 45. Newberry | Newberry | 34.30 | −81.62 | 145.1 | C00386209 | P | 1,227 ± 14 | 1,134 ± 11 |
| 46. Laurens | Laurens | 34.50 | −82.02 | 179.5 | C00385017 | P | 1,221 ± 15 | 1,131 ± 12 |
| 47. Ninety-Nine Islands | Blacksburg | 35.03 | −81.49 | 152.4 | C00386293 | P | 1,170 ± 11 | 1,094 ± 11 |
| 48. Saluda | Saluda | 34.00 | −81.77 | 146.3 | C00387631 | P | 1,254 ± 11 | 1,147 ± 11 |
| 49. Spartanburg 3SSE | Spartanburg | 34.91 | −81.91 | 185.9 | C00388188 | P | 1,243 ± 11 | 1,118 ± 11 |
| 50. Union 8Sᵉ | Union | 34.61 | −81.66 | 146.3 | C00388786 | P | 1,233 ± 14 | 1,118 ± 11 |
| 51. Wateree Dam | Lugoff | 34.33 | −80.70 | 70.1 | C00388979 | P | 1,244 ± 11 | 1,143 ± 10 |
| 52. Johnston 4SW | Johnston | 33.78 | −81.85 | 189 | C00384607 | P | 1,303 ± 15 | 1,160 ± 12 |
| 53. Clemson Oconee Co. AP | Clemson | 34.67 | −82.89 | 271.6 | W00053850 | P | 1,142 ± 14 | 1,171 ± 11 |
| 54. Greenwood Co. APᵈ | Greenwood | 34.25 | −82.16 | 192.3 | W00053874 | P | 1,208 ± 12 | 1,155 ± 11 |
| 55. Chesnee 7WSW | Chesnee | 35.11 | −81.97 | 228 | C00381625 | P | 1,211 ± 14 | 1,100 ± 9 |
| **Mean** | | | | | | P | **1,202 ± 11** | **1,137 ± 11** |
| 56. Caesars Head | Caesars Head | 35.11 | −82.63 | 975.4 | C00381256 | M | 986 ± 14 | 1,012 ± 12 |
| 57. Long Creek | Long Creek | 34.80 | −83.27 | 502.9 | C00385278 | M | 1,095 ± 11 | 1,083 ± 12 |
| 58. Pickens | Pickens | 34.88 | −82.72 | 354.2 | C00386831 | M | 1,176 ± 11 | 1,121 ± 10 |
| 59. Walhalla | Walhalla | 34.75 | −83.08 | 298.7 | C00388887 | M | 1,201 ± 14 | 1,113 ± 11 |
| **Mean** | | | | | | M** | **1,115 ± 10** | **1,082 ± 11** |

*Note.* Mean Hargreaves-Samani (H-S) and Priestley-Taylor (P-T) potential evapotranspiration (PET) for 18 years (1998–2015). Lat = latitude; Long = longitude; Elev = elevation; Reg = region; Co. = County; AP = airport; MU = municipal, BE = beach; ST = state; PA = park; Met. = metropolitan; D = downtown.
[a] Stations with pan evaporation and Penman–Monteith (P-M) PET. [b] H-S PET for 1996-2015. [c] P-T PET for 1998-2015. [d] Stations with P-M PET. [e] Stations with pan evaporation. [f] Data for Santee 2006–2015.
* Coastal P-T mean PET value significantly greater than for the Piedmont and Mountain. **Mountain (M) mean PET values significantly lower (α = 0.05) than values for Coastal (C) and Piedmont (P) regions for both methods. *

latitudes and longitudes were 32.48°–34.73°, 33.66°–35.11°, and 34.75°–35.11° and 78.72°–81.33°, 80.70°–82.89°, and 82.63°–83.27°, respectively (Table 1). Based on the NOAA site information, there were 6 stations near airports, 1 near a university, and 4 near water bodies in the Coastal region and 5 stations near airports, 2 near universities, and 3 near water bodies in the Piedmont region.

## PAN EVAPORATION MEASUREMENTS AND ANALYSIS FOR OPEN WATER PAN EVAPORATION

We obtained Class A PE data from the National Climatic Data Center's Global Historical Climatology Network website (https://www.ncdc.noaa.gov/cdo-web/search) for only 7 stations within the Coastal and Piedmont regions (no data for the Mountain) in South Carolina. The years for PE data varied from 1948 to 2014. In some cases data had gaps preventing us from calculating the annual values. The report by CDM Smith (2016) made available to us by South Carolina Department of Natural Resources also used data from the same site. For

water reservoirs like shallow lakes, OWE was calculated by multiplying measured PE data by pan coefficients (Jensen & Allen, 2016; Singh, 2016) obtained from NOAA Technical Report NWS-33 (Farnsworth et al., 1982) to the raw PE data. Annual pan coefficients typically range from 0.65 to 0.85, whereas monthly values can vary from 0.3 to 1.7, depending on water body characteristics, such as depth, turbidity, and potential for heat storage (Singh, 2016). The values unique for each station varying from 0.72 to 0.76 as reported by CDM Smith (2016) were also used in our study. These values closely agree with those developed by Phillips et al. (2016) for individual lakes due to their geographical and geometric (in shape and size) differences at the Savannah River site in South Carolina, except for the Coastal Plain region, with as high as 0.77 (McCuen, 1989). Although pan coefficients may bring uncertainties in estimates of OWE while calibrating PET estimates, the PE method is the only one that represents the measured evaporative demand for the study sites (McMahon et al., 2013).

## MODIFIED HARGREAVES-SAMANI (H-S; HARGREAVES & SAMANI, 1985) PET METHOD

Daily H-S PET was calculated using Eq. 1:

$$H-S\ PET\ (mm\ d^{-1}) = \frac{[0.0135(KT)(R_a)(T_{av}+17.8)(T_{max}-T_{min})^{0.5}]}{\lambda}, \quad (1)$$

where $T_{av}$, $T_{min}$, $T_{max}$, and $\lambda$ represent daily average, minimum, and maximum temperatures in °C, and factor for converting radiant energy flux in MJ m² d⁻¹ to mm d⁻¹, respectively. The term $((KT)(R_a)(T_{max}-T_{min})^{0.5})$ in Eq. 1 represents daily solar radiation ($R_s$) as suggested by Hargreaves-Samani (1982)

$$R_s = (KT)(R_a)(TD^{0.5}), \quad (2)$$

where $R_a$ is daily extraterrestrial radiation (MJ m⁻² d⁻¹) calculated using standard formulas following Allen et al. (1998) for given latitude and Julian day, and the term $(T_{max}-T_{min})$ is the temperature difference (TD). Hargreaves and Samani (1985) also provided the following field-based empirically calibrated equation for calculating KT:

$$KT = 0.00185\ (TD)^2 - 0.0433\ (TD) + 0.4023;$$
$$(R^2 = 0.70, SE = 0.0126). \quad (3)$$

In this study, KT, the empirical radiation adjustment factor, was calculated through trial and error by minimizing the average daily error obtained as a difference between the calculated daily solar radiation (using the assumed KT) and the actual measured daily solar radiation for a 5-year (2005–2009) record. Daily maximum and minimum temperature data for the same period were used for TD in Eq. 2. The estimated KT for the ten stations (5 from Coastal and 5 from Piedmont) ranged from 0.148 to 0.171; too few stations were available to perform this function for the Mountain region. After identifying no significant difference (α = 0.05) between the Coastal and Piedmont mean KT values (mean KT for Coastal: 0.157, very similar to the values [0.15–0.16] obtained by Amatya et al. (2000) for 3 coastal North Carolina sites using the self-calibration approach recommended by Allen, 1997; mean KT for Piedmont: 0.154), we used the average value (0.155) to compute adjusted H-S daily PET using Eq. 1 for each of the 59 stations. A mean value of KT = 0.154 was obtained in the H-S PET estimates for humid areas of Iran (Raziei & Periera, 2013).

## PRIESTLEY-TAYLOR (P-T; PRIESTLEY & TAYLOR, 1972) METHOD

Daily P-T PET was calculated using Eq. 4:

$$P-T\ PET\ (mm\ d^{-1}) = \left[1.26\left(\frac{\Delta(R_n-G)}{\Delta+\gamma}\right)\right]/\lambda, \quad (4)$$

where $R_n$ and G represent daily net radiation (MJ m⁻² d⁻¹) and daily soil heat flux (MJ m⁻² d⁻¹), respectively, and γ is psychrometric constant (kPa °C⁻¹) = (specific heat of air × atmospheric pressure)/(0.622 × λ), where atmospheric pressure (kPa) = 101.3 – 0.01055 × elevation (m). The parameters λ is latent heat of vaporization (MJ kg⁻¹) = 2.501 – (0.002361 × T) and Δ is slope of vapor pressure-temperature curve [kPa°C⁻¹] = 0.2 × (0.00738 × T + 0.8072)⁷ - 0.000116, where T = daily average air temperature (°C). The constant 1.26 is a calibration factor that accounts for aerodynamic effects for wet or humid conditions.

## PENMAN-MONTEITH (P-M; MONTEITH, 1965) MODIFIED BY ALLEN ET AL. (1998) AS FAO-56 METHOD

Daily P-M PET was calculated using Eq. 5 at stations where full data were available:

$$P-M\ PET(mm\ d^{-1}) = \frac{0.408\ \Delta(R_n-G)+\gamma\left(\frac{C_n}{T+273}\right)u2(e_s-e)}{\Delta+\gamma(1+C_d u2)}, \quad (5)$$

where, $R_n$, G, T, u2, and $(e_s-e)$ represent net radiation (MJ m⁻² d⁻¹), soil heat flux (MJ m⁻² d⁻¹), daily average temperature (°C), wind speed (m s⁻¹) adjusted to 2-m height above ground, and vapor pressure deficit (kPa), respectively. The psychrometric constant (g), slope of vapor pressure-temperature curve (Δ), numerator constant for reference type and calculation time step ($C_n$), and denominator constant for reference type calculation time step ($C_d$) were 0.0583 kPa C⁻¹, 0.1501 kPa C⁻¹, 900, and 0.34, respectively. Soil heat flux on a daily basis was estimated to be negligible in both the P-T and P-M methods (Allen et al., 1998; Amatya & Harrison, 2016).

All the daily PET values using H-S, P-T, and P-M methods were calculated directly in MS Excel spreadsheets using these equations with downloaded daily weather variables described next for multiple stations for the 1998–2015 period for the first 2 methods and 2002–2009 for the P-M method.

## WEATHER DATA ACQUISITION FOR PET ESTIMATES

**Air temperature (T).** The daily temperature (minimum and maximum) data for all stations, except for Santee (https://www.srs.fs.usda.gov/charleston/santee/), were downloaded from the NOAA website (https://www.ncdc.noaa.gov/cdo-web/search) and used to obtain the daily average temperature to compute daily H-S PET using Eq. 1 for the 1998–2015 period. In the case of a missing value between dates, the average of the previous and subsequent daily temperature value was used. For situations where more than 1 consecutive date had missing values, 1-km gridded temperature data downloaded from the Oak Ridge National Laboratory's DayMet website (https://daymet.ornl.gov/singlepixel) were used to fill the data gap because of its good agreement with measured data ($R^2 > 0.90$) and a near unity slope (0.97 to 1.03) with a small bias for randomly selected 4 stations (2 from Piedmont and 2 from Coastal).

**Solar radiation ($R_s$).** Daily solar radiation ($R_s$) data were used for calculation of net radiation for P-T (Eq. 4) and P-M (Eq. 5) PET methods because measured net radiation data were only available for the Santee Station (Coastal region). Actual field measured solar radiation data were available for only a very few Coastal stations (Santee, Savannah River site, and North Inlet); therefore, we used the National Solar Radiation

(NSR) database website (https://nsrdb.nrel.gov) to download modeled hourly solar radiation data on 10-km grid for all 59 stations for the period 1998–2015. The hourly data were processed to compute daily values. For this study, we did not directly use the daily solar radiation obtained from these modeled hourly measurements for computing net radiation for the P-T and P-M PET estimates. We first selected stations with actual measured daily solar radiation and regressed with daily data obtained from the downloaded hourly data for all of those selected stations to obtain a combined calibration equation that was used for all stations without measured data. Because the calibration Eq. 6 obtained by a bootstrap regression (Ssegane et al., 2017; that potentially reduces the effects of autocorrelation in daily values) was significant, we chose it to correct the bias in daily solar radiation data (Mj m$^{-2}$ d$^{-1}$) from the downloaded hourly data for all the stations.

$$R_s = \text{NSR data} \times 0.98 - 1.59$$
$$(R^2 = 0.88; P < 0.001;$$
$$\text{RMSE} = 1.89 \text{ Mj m}^{-2}\text{ d}^{-1}), \tag{6}$$

where, RMSE = root mean square error.

**Net radiation ($R_n$).** Daily net radiation (Mj m$^{-2}$ d$^{-1}$) needed for computing the daily P-T and P-M PET (Eqs. 4 and 5) was calculated following the FAO (1990) and Archibald and Walter (2014) methods based on modeled (where measured was not available) daily solar radiation and temperature:

$$R_n = (1 - \alpha) R_s - C \varepsilon_n \sigma T_k^4, \tag{7}$$

where, $\alpha$ is albedo ( = 0.23 for grass), C is cloudiness factor, $\varepsilon_n$ is net emissivity ( = atmospheric emissivity – vegetation emissivity), $\sigma$ is Stephan–Boltzman constant = $4.89E^{-09}$ Mj m$^{-2}$K$^{-4}$d$^{-1}$, and $T_k$ is temperature in Kelvin ( = T °C + 237.3).

One slight modification we made was in computing the cloudiness factor (C):

$$C = (a_c \times R_s / R_a + b_c), \tag{8}$$

in which the constants $a_c$ and $b_c$ recommended in the FAO (1990) method were optimized to 0.72 and 0.28, respectively, by minimizing the difference between daily calculated net radiation by the FAO method and the measured data from the USDA Forest Service SEF Station for a 10-year (2006–2015) period. This maximized the highest Nash–Sutcliffe coefficient (in this case, N-S = 0.93). The calibrated constants were used to compute cloudiness as required for calculating net radiation at each station. Measured net radiation data were available for only the Coastal SEF station starting from 2006 to 2015. Daily calculated net radiation using the FAO (1990) method yielded a strong significant relationship ($R^2 = 0.94$; $P = 0.02$; RMSE = 1.21 Mj m$^{-2}$ d$^{-1}$) obtained using a block bootstrap geometric linear regression with the measured data at the SEF station. The $\varepsilon_n$, net emissivity was calculated from the Eq. 9:

$$\varepsilon_n = 0.261 \text{ EXP}(-7.77 \times 10^{-4} \text{ T}^2) - 0.02. \tag{9}$$

**Relative humidity and wind speed.** Measured hourly wind speed (u) and relative humidity (RH) data were available only for 11 stations (Charleston Airport [AP], North Myrtle Beach AP, Florence AP, Myrtle Beach Air Force Base, Beaufort Marine Corps Air Station, Darlington County AP, Columbia Metropolitan AP, Columbia Owens AP, Greenville Downtown AP, Greenwood County AP, and Greenville–Spartanburg AP) from NOAA's NCEI website (https://www.ncdc.noaa.gov/cdo). Missing data were gap filled using the same procedure described for the air temperature. Hourly values were further processed to obtain the daily values. Daily humidity values were used for calculation of vapor pressure deficit (as shown next), which together with the daily wind speed was used in calculating P-M PET in Eq. 5.

**Vapor pressure deficit.** The vapor pressure deficit was calculated using the maximum ($T_{max}$) and minimum ($T_{min}$) temperature and maximum ($RH_{max}$) and minimum ($RH_{min}$) relative humidity as follows following the procedure by Jensen et al. (1990):

Maximum vapor pressure, $e_o(T_{max}) = \text{EXP}[(16.78 \times T_{max} - 116.9) / (T_{max} + 273.15)]$;

Minimum vapor pressure, $e_o(T_{min}) = \text{EXP}[(16.78 \times T_{min} - 116.9) / (T_{min} + 273.15)]$;

Saturated vapor pressure, $e_s = [e_o(T_{max}) + e_o(T_{min})] / 2$;

Actual vapor pressure, $e_a = 0.5 \times \{[RH_{min} / 100 \times e_o(T_{max})] + [RH_{max} / 100 \times e_o(T_{min})]\}$;

Vapor pressure deficit (VPDC) = $e_s - e_a$ (10)

## DATA PROCESSING AND STATISTICAL ANALYSES

First, we used the daily weather variables (temperature, humidity, and solar radiation) to compute daily net radiation (Eq. 7) and vapor pressure deficit (Eq. 10). As a next step, we used daily weather data to calculate daily PET by each of the prior 3 methods (Eqs. 1, 4, and 5), although PET estimates from weather data at the monthly scale have been reported to be acceptable by some studies for assessing water yield of stream or river basins (CDM Smith 2016; Hember et al., 2017; Lu et al., 2003; Rao et al., 2011; Shevenell, 1996). Daily PET values were integrated to obtain the monthly and annual means for each of the stations in each of the 3 regions. Then the annual means were averaged to obtain the mean annual value, and similarly, monthly means for each year were averaged to obtain the mean monthly value at each of the stations. Regional mean monthly and mean annual values by each of the PET methods were derived by averaging station mean values in each of the 3 regions. The same procedure was repeated for the OWE data obtained from the PE stations to summarize mean monthly and mean annual values. Standard deviations and standard errors were also reported.

The significance in differences in mean annual PET among the regions by each method (H-S, P-T, and P-M) and among the methods in each region were identified using analysis of variance with Tukey test (α = 0.05) in R software (R Core Team, 2017). For example, we tested whether there was a difference among mean annual P-T PET in Coastal, Piedmont, and Mountain regions and also whether there was a difference in mean annual PET by H-S, P-T, and P-M methods in Piedmont region. Regressions between H-S PET (1996–2015), P-T PET (1998–2015), and P-M PET (2002–2009) were developed to identify the strength of relationships between each pair, particularly for the H-S and P-T PET with the standardized P-M PET.

Scatter plots and ordinary least squares lines were fitted to examine the association between the monthly PE and monthly PET by each of the 3 methods for all regions together. Unlike with daily values, it was assumed that monthly model relationship would have a negligible effect of autocorrelation. Statistical criteria of coefficient of determination ($R^2$), Nash–Sutcliffe efficiency (NSE), percent bias (PBIAS), and root mean square error normalized by dividing by standard deviation (RSR) following Moriasi et al. (2007) were used to evaluate the model performance.

### GIS SPATIAL ANALYSES FOR INTERPOLATED PET MAPS

The above analyses provided seasonal and annual estimates of PET for 59 stations spread across South Carolina (Figure 1). However, these stations may not yet adequately describe the PET estimates for some specific sites of interest due to their location which may be much further away than desired. In such circumstances an accurate spatial interpolation of the PET from the spatially distributed station data is needed, and the method of interpolation between plays an important role, as with any other hydrologic data (Chen et al., 2017).

Among several interpolation methods available in the literature, IDW surface interpolation scheme available in ArcGIS 10.5 tool was used to develop spatial PET distribution maps for South Carolina. IDW uses the influence of a measured point by weighing it according to the distance from the sampled point (in our case, the PET values of NOAA weather stations) to the estimated point.

The mean monthly and mean annual PET calculated by P-T and H-S methods for all 59 NOAA weather stations in the state were added to the weather station shape file. The IDW scheme created the spatial distribution raster, which was masked to the state boundary and classified into 5 class ranges to provide the visual distributed map of PET in the state. An automated geospatial model was developed in ArcGIS ModelBuilder to streamline the entire working process and make the job efficient.

More details on the weather stations, metadata, data gap, data filling and extrapolation, relationships of climatic data with PE, calculation of daily PET by each method and

their parameters, statistical methods used for PET model evaluation, and GIS maps for the mean annual and mean monthly PET by both the H-S and P-T PET methods are given in the Final report of this project being submitted to South Carolina Department of Natural Resources.

## RESULTS AND DISCUSSION

### SPATIAL AND TEMPORAL DISTRIBUTION OF PET

The 18-year (1998–2015) mean annual PET with their standard errors of the mean using H-S and P-T methods are presented in Table 1 for stations in each of the 3 regions, with year-to-year variability shown in Figures 2A and 2B. However, a comparison among all the 3 methods was also done for 11 stations in common in the Coastal and Piedmont regions (Table 2) for an 8-year (2002–2009) period for which complete data for the P-M method were available.

The highest mean annual PET was computed for the P-M method followed by the H-S PET and the P-T PET for both the Coastal and Piedmont regions. The difference in mean annual means between the H-S PET and P-T PET was less than 2% compared with more than 7% between the P-T and the P-M methods.

When the mean annual PET values were correlated to elevation across regions, both the H-S and P-T PET significantly ($P < 0.05$) decreased with increasing elevation from Coastal to Mountain (not shown), consistent with findings in Shevenell (1996). The mean annual H-S PET for Piedmont and Mountain regions significantly decreased ($P < 0.05$) with increase in elevation unlike the Coastal region with only a small gradient (not shown). The mean annual PET trends for the regions also followed the temperature and net radiation data (not shown); the increase of which correlated to an increase of PET.

**Comparison of 3 PET methods in each region.** Calculated annual mean PET for each of the 3 regions varied as high as about 1,300 mm for the H-S PET and 1,250 mm for the P-T PET for the Coastal region to as low as 1,000 mm for the P-T method to 1,020 mm for the H-S PET in the Mountain region (Figures 2A and 2B). Clearly, annual mean H-S PET was higher than the P-T PET in each of the 3 regions. Although both methods yielded similar annual trend, the high and low PET values did not necessarily coincide for all the years. Furthermore, the difference in Coastal and Piedmont regions was smaller for the H-S method compared with the P-T method. This was attributed to the effects of only temperature on the H-S and both the temperature and radiation in the P-T method.

The mean annual H-S PET for Coastal, Piedmont, and Mountain regions were 1,232 mm, 1,202 mm, and 1,115 mm, respectively, compared with 1,179 mm, 1,137 mm, and 1,082 mm, respectively, for the P-T PET (Table 1).

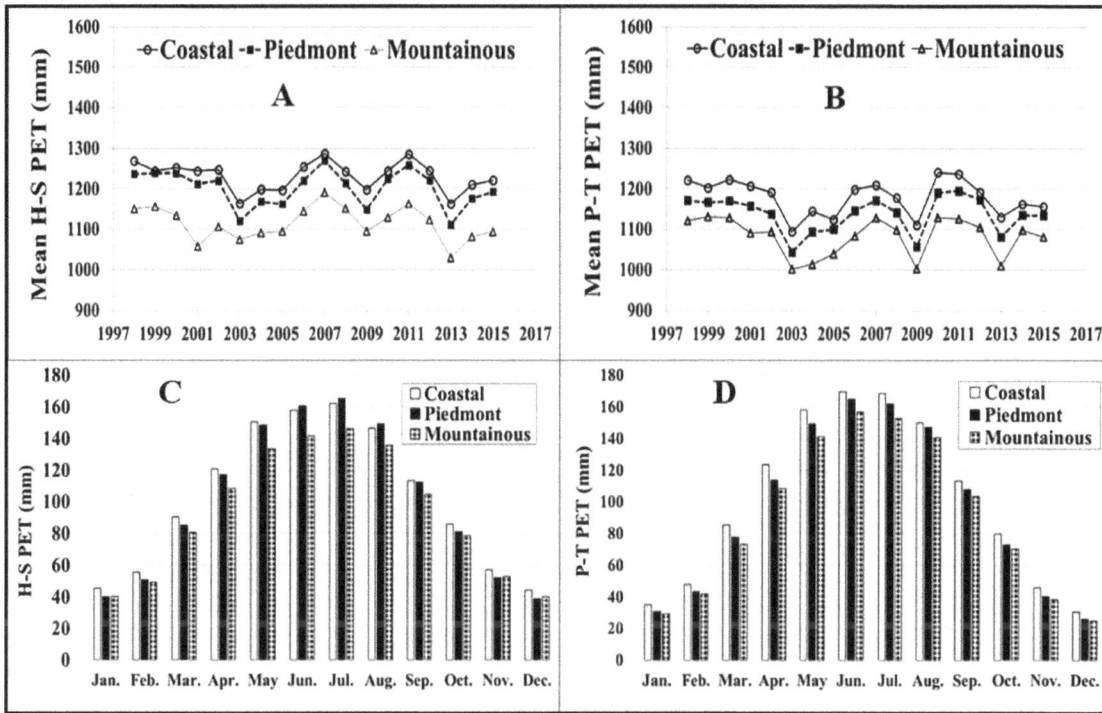

**Figure 2.** Annual mean and monthly mean (A and C) Hargreaves–Samani (H-S) potential evapotranspiration (PET) and (B and D) Priestley–Taylor (P-T) PET for the Coastal, Piedmont, and Mountain regions.

**Table 2.** Comparison of mean annual Hargreaves–Samani (H-S), Priestly–Taylor (P-T), and Penman–Monteith (P-M) potential evapotranspiration (PET; in mm).

| Station | Reg | H-S PETa ± SE (COV) | P-T PETa ± SE (COV) | P-M PET ± SE (COV) |
|---|---|---|---|---|
| Charleston AP | C | 1,197 ± 10 (0.02) | 1,178 ± 15 (0.04) | 1,270 ± 25 (0.05) |
| North Myrtle Beach AP | C | 1,044 ± 11 (0.03) | 1,128 ± 15 (0.04) | 1,165 ± 23 (0.05) |
| Florence Airport | C | 1,187 ± 14 (0.03) | 1,142 ± 17 (0.04) | 1,244 ± 26 (0.06) |
| Myrtle Beach AFB | C | 1,091 ± 19 (0.05) | 1,164 ± 17 (0.04) | 1,248 ± 24 (0.05) |
| Beaufort MCAS | C | 1,181 ± 15 (0.04) | 1,214 ± 16 (0.04) | 1,239 ± 22 (0.05) |
| Darlington Co. AP | C | 1,250 ± 20 (0.04) | 1,147 ± 16 (0.04) | 1,267 ± 36 (0.06) |
| Columbia Met. AP | C | 1,207 ± 15 (0.04) | 1,139 ± 16 (0.04) | 1,252 ± 26 (0.06) |
| Columbia Owens AP | C | 1,203 ± 15 (0.03) | 1,152 ± 19 (0.05) | 1,149 ± 34 (0.07) |
| Average for C | | 1,170 | 1,158 | 1,229 |
| Greenville D AP | P | 1,107 ± 16 (0.04) | 1,125 ± 18 (0.04) | 1,142 ± 31 (0.07) |
| Greenwood Co. AP | P | 1,192 ± 20 (0.05) | 1,127 ± 16 (0.04) | 1,240 ± 41 (0.07) |
| Greenville–Spartanburg AP | P | 1,130 ± 20 (0.05) | 1,109 ± 20 (0.05) | 1,213 ± 38 (0.09) |
| Average for P | | 1,143 | 1,120 | 1,198 |

*Note.* Reg = region; COV = coefficient of variation; AP = airport; AFB = Air Force Base; MCAS = Marine Corps Air Station; Met. = metropolitan; C = Coastal; D = downtown; P = Piedmont. Mean annual PET for the period, 2002–2009. No Mountain weather station had more than temperature data, preventing the multiple-model comparison.
[a] Significantly different (α = 0.05) from P-M PET.

Mean annual H-S PET was significantly ($P < 0.05$) greater than P-T PET for all 3 regions. Mean annual H-S PET values varied from 1,062 mm for North Myrtle Beach to 1,409 m for Yemassee in southern Coastal South Carolina, with a mean of 1,232 mm for the Coastal region (Table 1); however, the P-T values had a much smaller variability in mean annual PET with 1,132 mm for Santee to 1,235 mm for Beaufort, with a mean of 1,179 mm. In the Piedmont H-S PET varied from 1,093 mm in Greenville to 1,324 mm in Columbia, with a mean of 1,202 mm, whereas the P-T PET ranged from 1,094 mm at Ninety Nine Islands to 1,203 mm in Columbia, with a mean of 1,137 mm. Similar observations were found with results from the 2 methods for the limited 4 stations in the Mountain, with the lowest at Caesar Head by both the methods to highest at Walhalla by the H-S method and at Pickens by the P-T method. Perhaps because of interaction of both the temperature and net radiation the P-T method did not necessarily yield higher PET for the southern stations than the north in each region (Table 1). Annual H-S PET also had a larger coefficient of variance than the P-T method at all stations. This was likely due to wider variation in air temperatures between stations from the south to the north (reflected in the H-S PET estimates) compared with the solar radiation (used in the P-T PET method) that did not vary as much. Mean annual P-T PET was lower than the H-S PET for the 18-year period (Table 1).

When compared among all 3 methods using only the short 8-year data (Table 2), P-M method varied from 1,142 mm at the Piedmont station to 1,270 mm at the Coastal. Similar pattern was observed for P-T from 1,109 mm at Piedmont to 1,214 mm at Coastal. However, it did not hold for the temperature only based H-S method, with both the lowest and highest PET occurring at the Coastal stations. P-M mean annual PET was the highest for all stations in the 2 regions, except for the Columbia Owens station in the Coastal where H-S PET yielded the highest. The H-S PET and P-T PET for Coastal and Piedmont were significantly ($P < 0.05$) lower than P-M PET (Table 2). The coefficients of variation were the highest for the fully process-based P-M method with multiple variables followed by the P-T and H-S method with only the temperature variable.

Monthly mean PET for the H-S and P-T PET for each of the 3 regions are presented in Figures 2C and 2D, respectively. The monthly mean P-T PET consistently yielded highest values for the Coastal followed by the Piedmont and Mountain regions, unlike the H-S PET which did not indicate any pattern. Again this may likely due to both decreasing temperature and radiation from Coastal to Mountain.

The mean monthly H-S PET for Coastal, Piedmont, and Mountain regions were 102 mm, 101 mm and 94 mm, respectively, which were very similar to the mean monthly P-T PET of 102 mm, 95 mm and 91 mm, respectively, for the 3 regions. The highest monthly mean PET for both the H-S and P-T methods (as high as 171 mm for the P-T in the Coastal) were observed in June and July and the lowest PET in December (as low as 25 mm for the P-T method in the Mountain) in each of the 3 regions (Figure 4).

Interestingly, the temperature-based H-S monthly mean PET were higher than the P-T PET during the fall-winter and early spring in contrast with the P-T PET, which yielded higher PET than the H-S method during the May–August summer months when both the radiation and temperature included in the P-T method are generally higher than the rest of the months. Garcia et al. (2004) reported large variations of energy between summer and winter with the greatest radiation energy occurring in summer in Bolivian highlands. Sumner et al. (2017) analyzed REF-ET data for stations in Florida, and they reported that monthly ET peaked in June and July and that the greatest variabilities were observed in spring and summer, with less variability for the winter months. Amatya et al. (1995) reported for a site in eastern North Carolina that peak P-T PET values occurred in summer (June and July) and that those peak values were in close agreement with results from P-M PET. Amatya and Harrison (2016) documented that the peak monthly PET occurred in June and July at the SEF weather station, and were associated with the peak values of the energy component and leaf area index at the same period.

The observed differences in calculated PET by these 3 different methods for both the mean monthly and annual periods was likely due to different variables used in the methods from process based P-M method that uses all variables like temperature, net radiation, wind speed and relative humidity to somewhat simpler temperature only based H-S method. In cooler months, wind speed is likely to be more important than solar radiation as opposed to the solar radiation in the summer months, potentially indicating importance of the P-M method. It is also important to note that only the H-S method uses complete measured data (temperature) in contrast with the P-T and P-M methods that also use modeled radiation data potentially increasing some uncertainty.

**Regional comparison of PET by each method.** The mean annual H-S PET of 1,115 mm for Mountain region was significantly ($P < 0.05$) lower than for the Coastal (1,232 mm) and Piedmont (1,202 mm) regions (Table 1), likely due to the lowest temperatures observed for the Mountain region. There was no significant difference in H-S PET between Coastal and Piedmont regions (Figure 2A). The mean annual P-T PET of 1082 mm for Mountain region was significantly ($P < 0.05$) lower than for the Coastal (1,179 mm) and Piedmont (1,137 mm; Figure 2B), also likely due to both significantly lower temperature and slightly lower net radiation (not shown).

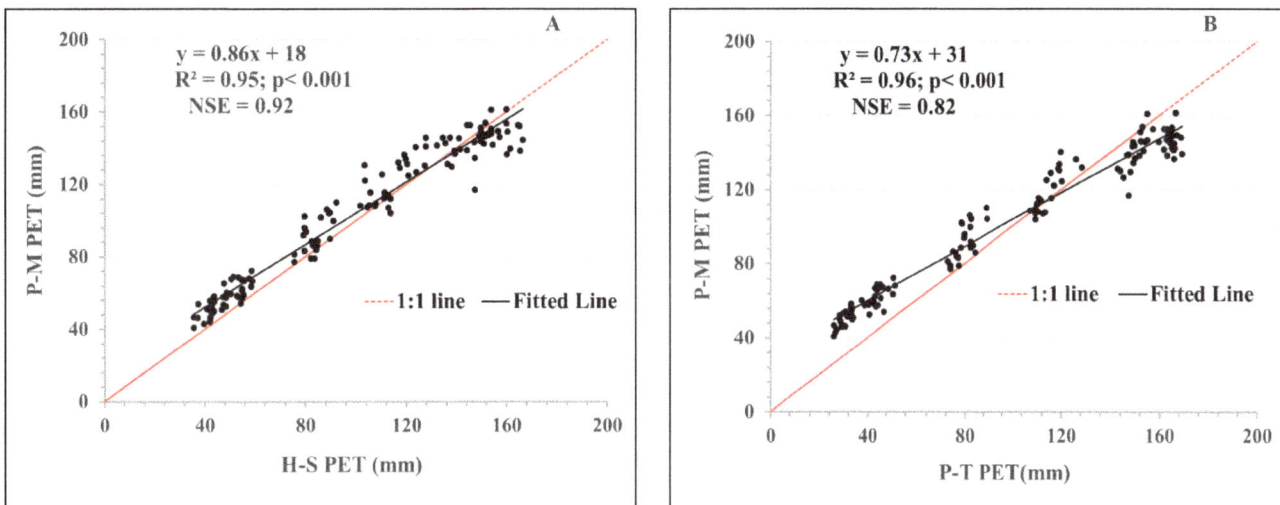

**Figure 3.** Relationships between monthly (A) Penman–Monteith (P-M) and Hargreaves–Samani (H-S) potential evapotranspiration (PET) and (B) P-M and Priestley–Taylor (P-T) PET.

The P-T PET for Coastal region was significantly ($P < 0.05$) greater than for the Piedmont region (Figure 2B).

The mean annual P-M PET (1,229 mm), obtained using data from only limited stations (Table 2), for the Coastal region was not significantly ($P > 0.05$) different from the Piedmont region (1,198 mm), likely due to no significant differences in weather variables (not shown). However, the H-S PET and P-T PET for both the Coastal and Piedmont were significantly ($P < 0.05$) lower than the P-M PET based on data from these limited stations (Table 2).

Our estimates of the mean annual standard P-M PET (mean annual for Coastal: 1,229 mm; mean annual for Piedmont: 1,198 mm) are somewhat smaller (within 6%) than those recently reported by NRCS (2016; mean annual for Coastal: 1,266 mm; mean annual for Piedmont: 1,270 mm) using data from 2002 to 2009 for 14 stations (out of the 59 in this study). However, we could not verify the source of weather variables the NRCS study used in their P-M PET estimate for those stations. Future study should verify results from these 2 studies using the same P-M PET method for those 14 stations.

The regional differences in weather variables (e.g. sunshine hours, temperature, humidity, radiation, rainfall and cloud cover) during different seasons contributed most to the seasonal variability in H-S and P-T PET results (Figures 2C and 2D), consistent with past studies (Chattopadhyay & Hulme, 1997; Hember et al., 2017; Shukla & Mintz, 1982; Thomas, 2000). For example, increases in temperature led to increases in both the H-S PET and P-T PET in Coastal, Piedmont, and Mountain regions. Hember et al. (2017) documented that the increases and decreases of PET with weather variables like temperature are more pronounced at shorter time intervals. Barik et al. (2016) found greater PET values calculated during clear days.

Figure 3 shows the relationship between monthly H-S PET and P-M PET, as well as the relationship between the monthly P-T PET and the P-M PET using data from stations in both the Coastal and Piedmont regions (Table 2), as there was inadequate data with only 3 stations for the Piedmont region to analyze it separately. Both the slope and the NSE for the H-S PET method (0.86; 0.92) were higher than that for the P-T method (0.73; 0.82). Similarly, the PBIAS (–0.04%) for the H-S PET was lower than that for the P-T method (–0.16%). However, the RSR value was higher (0.22) for the H-S method compared with 0.19 for the P-T method. These evaluation statistics for performance of both the H-S and P-T PET methods compared with the standard P-M method for monthly PET estimates can clearly be rated as "very good" based on the Moriasi et al. (2017) recommended criteria of $0.00 < RSR < 0.50$, $0.75 < NSE < 1.00$, and $PBIAS < \pm 10$ for monthly streamflow estimates. Based on this evaluation and the fact that the mean monthly PET between the H-S and P-T PET were also similar, either of the method can be used.

## COMPARISON OF PET METHODS WITH PAN (PE) AND OPEN WATER EVAPORATION (OWE)

Examination of relationships between the measured mean monthly PE and the PET calculated using H-S and P-T with 84 observations from 7 stations and with P-M PET with only 24 observations from 2 stations where the data was available in Figure 4, yielded the $R^2$ values of 0.92, 0.94, and 0.97, respectively, with under predictions of PET (as much as by 28 mm for the P-M method) by all 3 methods as shown by their slopes >1. This shows that the calculated PET values by all 3 methods are somewhat realistic because evaporation measured from pans is generally greater than from nearby vegetated areas (Grismer et al., 2002; Shevenell, 1996). For instance, reference crop evaporation is typically lowered by multiplying the PE value by 0.65 to 0.85 for annual and 0.3

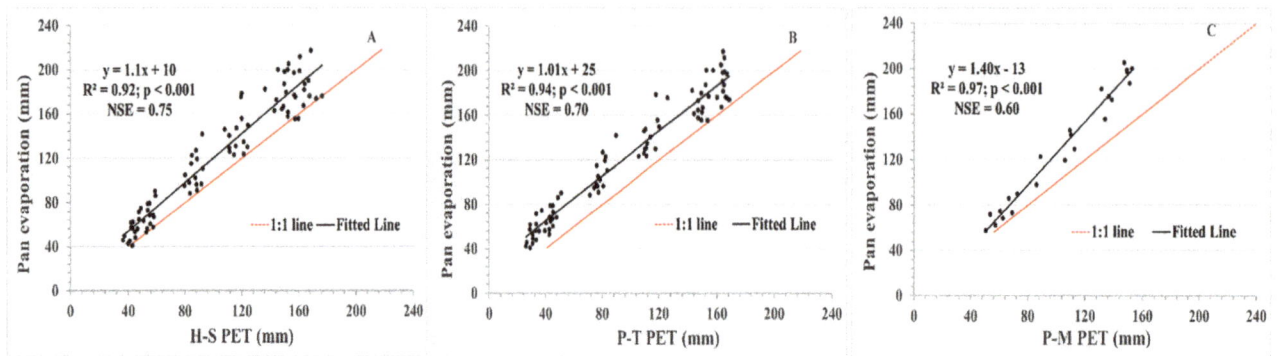

**Figure 4.** Relationships between mean monthly pan evaporation and (A) Hargreaves–Samani (H-S) potential evapotranspiration (PET), (B) Priestley–Taylor (P-T) PET using all 7 stations' data in Table 3, and (C) Penman–Monteith (P-M) PET using only 2 stations' data.

to 1.7 for monthly, depending on wind speed, and the fetch of wet versus dry crop (Maidment, 1993; Singh, 2016). The clustering of points, particularly in plots of Figure 4A and 4B, indicates the effects of seasonal climatic variables on PET by both the H-S and P-T methods. Based on the highest NSE of 0.75 and the smallest RSR of 0.21 and PB of 16.7% compared with 2 other methods, H-S method was found to be most closely associated with the mean monthly pan data. The P-T method was the second most closely associated (NSE = 0.70; RSR = 0.23, PB = 24.6%) followed by the P-M method (NSE = 0.60; NRMSE = 0.24; PB = 19.3%) using only the limited data. Because the H-S method generally overestimates the PET in the humid regions (Amatya et al., 1995; Dai et al., 2013) its close association with the pan data is expected. Flint and Childs (1991) reported α = 1.26 in the P-T method could represent OWE estimate for humid regions.

Mean annual PET estimated by each of the 3 methods was compared with the OWE obtained from the measured PE for 7 stations, with 2 in Coastal and 5 in the Piedmont (Table 3; Figure 1). The OWE values ranged from 1,095 mm to 1,272 mm for the Coastal and 970 mm to 1,071 mm for the Piedmont, indicating higher values for the Coastal than the Piedmont. This shows that the PET estimates discussed earlier by all 3 methods with higher values in Coastal than the Piedmont are consistent with the measured weather variables as well as the PE data used to obtain OWE values. The average annual percent deviations of H-S PET and P-T PET from OWE for the 7 stations were 13.5 and 6.6, respectively (Table 3). The smaller average deviation yielded by the P-T method is consistent with other studies (Rosenberry et al., 2007; Winter et al., 1995). Although the H-S PET had closer association with measured PE data, the higher mean percent deviation from the OWE was likely due to use of annual pan coefficients, varying from 0.72 to 0.76, obtained from the CDM Smith (2016) report. The estimated OWE were consistently lower than the PET estimated by either of the H-S or P-T methods in the region.

**Table 3.** Comparisons of mean annual open water evaporation (OWE) versus calculated potential evapotranspiration (PET), and percentage deviation of calculated value from corrected measured values.

| Station | Reg | OWE, mm | H-S PET, mm | P-T PET, mm | Deviation, % H-S PET | Deviation, % P-T PET |
|---------|-----|---------|-------------|-------------|----------------------|----------------------|
| Charleston | C | 1,271.5 | 1,201 | 1,205 | −6.6 | −5.5 |
| Florence | C | 1,094.7 | 1,206 | 1,168 | 10.2 | 6.3 |
| Sandhill | C | 1,207.0 | 1,234 | 1,166 | 2.3 | −3.5 |
| Clarks Hill | P | 969.8 | 1,290 | 1,174 | 33.0 | 17.4 |
| Clemson | P | 1,046.5 | 1,166 | 1,147 | 11.4 | 8.8 |
| Blackville | P | 1,070.6 | 1,260 | 1,182 | 17.7 | 9.4 |
| Union | P | 969.5 | 1,229 | 1,118 | 26.8 | 13.3 |
| Average | | 1,090 | 1,227 | 1,166 | 13.5 | 6.6 |

*Note.* Reg = region; H-S = Hargreaves–Samani; P-T = Priestley–Taylor; C = Coastal; P = Piedmont. Source for OWE: CDM Smith report 2016. Source for H-S PET (1998–2015): this study. Source for P-T PET (1998–2015): this study.

Some uncertainties exist in derived pan coefficients due to pan type (screened/unscreened), ground cover, evaporative conditions within the fetch of the pan, presence of plants and foreign materials, microclimatic conditions surrounding the pan (i.e., freezing), and the level of maintenance. Future study should evaluate mean monthly pan coefficients developed using climatic data at the stations (Grismer et al., 2002; Irmak et al., 2002) and estimate $ET_0$ using these coefficients with measured PE data to recompare them against $ET_0$ estimates obtained by the H-S and P-T PET methods.

## DISCUSSION ON H-S AND P-T PET METHODS

These results on the comparisons between the H-S and P-T methods in each region and comparisons of all 3 methods including the P-M with the measured PE lead us to conclude that the H-S PET method is most likely the best method followed by the radiation-based P-T for application in the state of South Carolina. However, Amatya et al. (1995) found the P-T method superior to H-S method, which consistently overpredicted PET by the standard P-M method for the North Carolina Coastal Plain site. However, when making management decisions about applying either method, it may also be important to consider some other factors including the data availability and quality and land use. The fact that the radiation data used to estimate P-T PET for 59 stations in this study were modeled and calibrated with data from a few Coastal stations might have also influenced its results. The land use/land cover in river basins of South Carolina is composed of about 60% forests, on average. Amatya and Harrison (2016) showed a closer agreement of the P-T PET with the forest-reference based P-M PET for a coastal forest in South Carolina. Similarly, Rao et al. (2011) found higher correlation between simulated and measured monthly streamflow using the P-T PET than even the P-M PET in their hydrologic model applied on upland forests in western North Carolina. Lu et al. (2005) also found the P-T method performing better than the H-S method for 36 forested watersheds in the southeast. Archibald and Walter (2014) also found a stronger correlation of the P-T method with measured ET during periods of maximal ET than the fully empirical Hargreaves, Hamon and Oudin methods. Furthermore, literature suggests that P-T method also performs better than the temperature-based methods in estimating OWE (McMahon et al., 2013; Rosenberry et al., 2007; Winter et al., 1995). However, a recent study by Amatya et al. (2016) also found a satisfactory performance in predicting monthly streamflow of a coastal forest watershed in South Carolina when the H-S based PET adjusted to match the P-M PET was applied to simulate ET in a hydrologic water balance model. One reason for a better agreement of

the H-S method with the P-M method in our study compared with other studies is likely due to adjustment of the original H-S method with the KT factor (0.155) calibrated based on measured solar radiation from 5 Coastal and 5 Piedmont stations and applied to all the stations. This is consistent with a recent study by Raziei and Periera (2013). Therefore, based on all these facts, we recommend using the P-T method for South Carolina if and when measured radiation data are available; otherwise, the H-S method adjusted for the KT parameter in this study should be adequate for monthly water balance, crop water requirements, and surface and groundwater modeling purposes. That said, both H-S and P-T PET could be used for these conditions with PE data to develop mean monthly correction factors for application in lake/OWE analyses.

The mean annual H-S and P-T method-based spatially distributed PETs are presented in Figure 5(left) and 5(right), respectively.

The monthly mean H-S and the P-T-based PET results for 59 stations within 3 regions shown in Figure 1 were used for creating spatially interpolated GIS-based monthly mean PET maps for the whole state of South Carolina, as shown in Figures 6 and 7, respectively. The greatest interpolated monthly PET values were observed in the Coastal region followed by Piedmont and smallest values or the Mountain region for both the methods. The interpolated mean monthly PET range was 40–47 mm for the month of January to 154–170 mm for July for the P-T method. The PET ranges increased from January and peaked in June and there after exhibited a decreasing trend to December. The variability in monthly ranges could be due to variability in both the temperature and net radiation. We also developed the maps of 95% confidence limits of the mean (not shown).

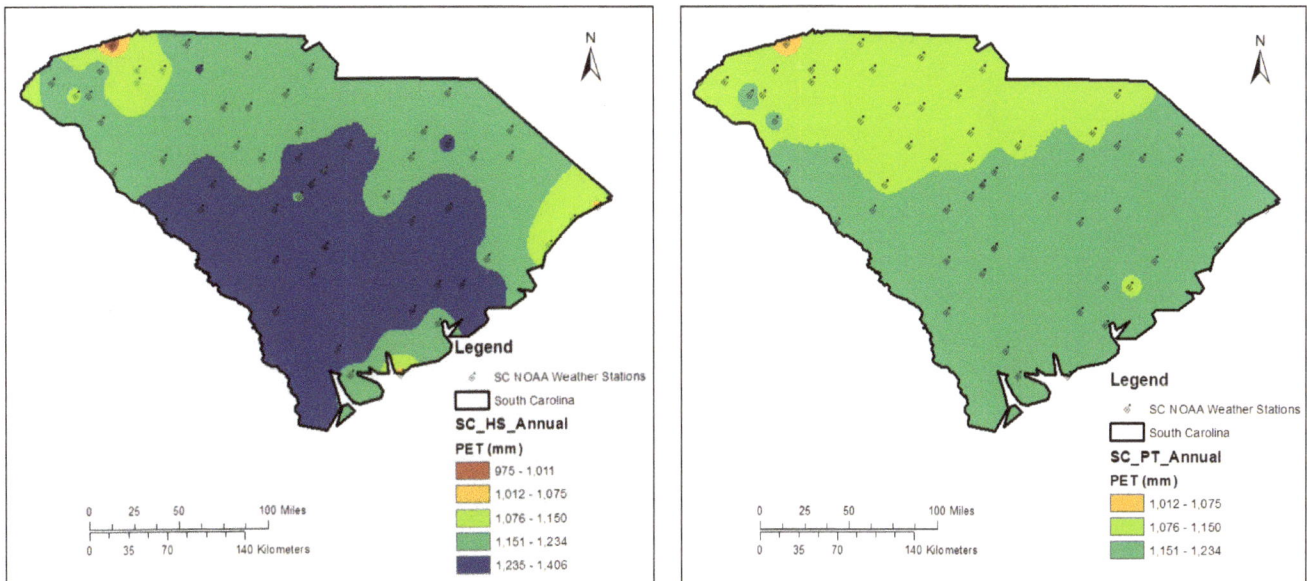

**Figure 5.** Interpolated mean annual Hargreaves-Samani (H-S) PET model for years 1996-2015 (left) and Priestly-Taylor (P-T) PET model for years 1998-2015 (right) determined from analysis of weather data at stations across South Carolina. Note the upper-most and lower-most ranges are not present for the P-T model because of the smaller variance in the results. The range-in-value bins are the same for both maps.

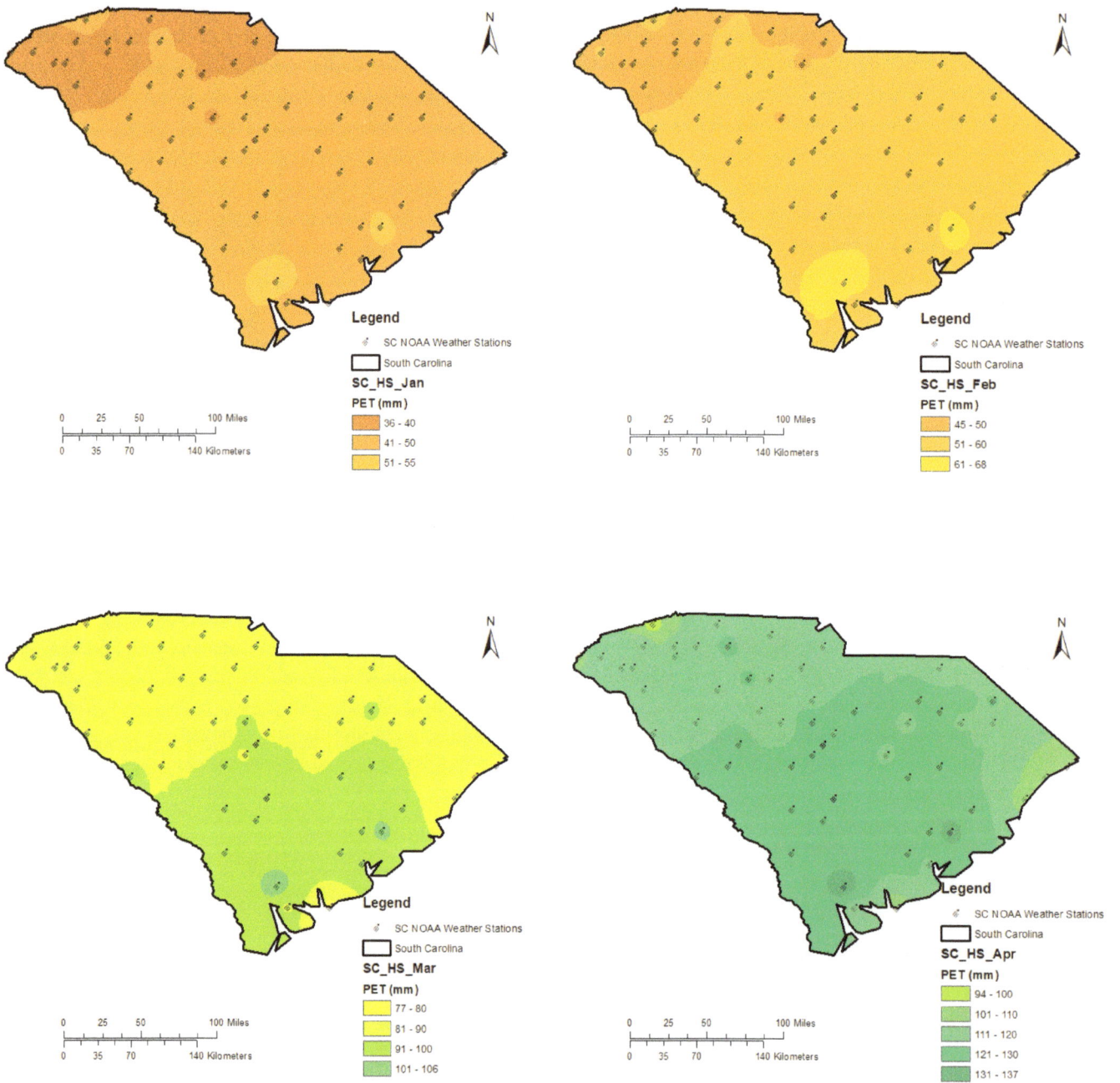

**Figure 6a.** Interpolated mean monthly H-S PET for January and February (top row) and March and April (bottom row). Data period of analysis is 1996-2015.

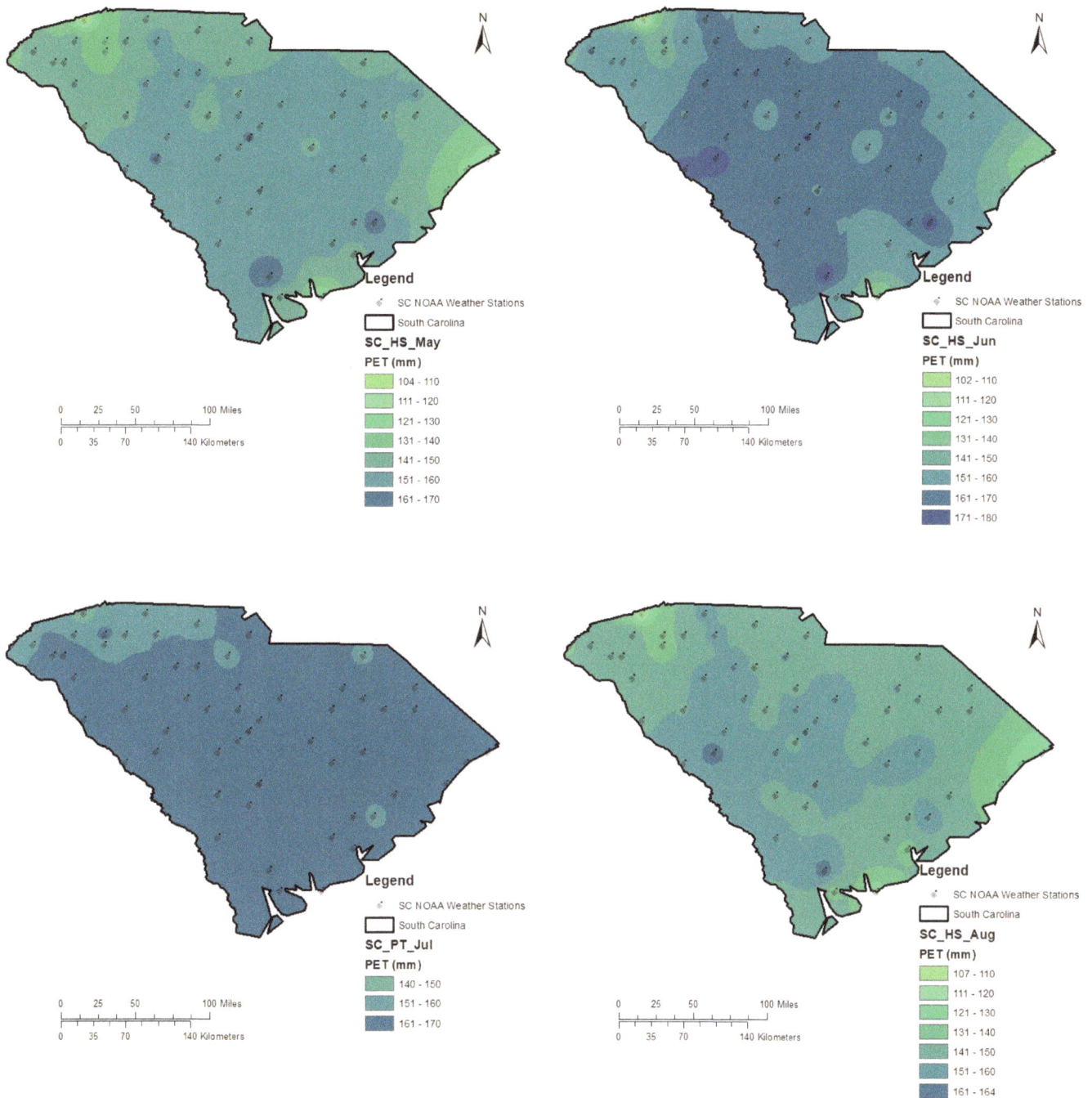

**Figure 6b.** Interpolated mean monthly H-S PET for May and June (top row) and July and August (bottom row). Data period of analysis is 1996-2015.

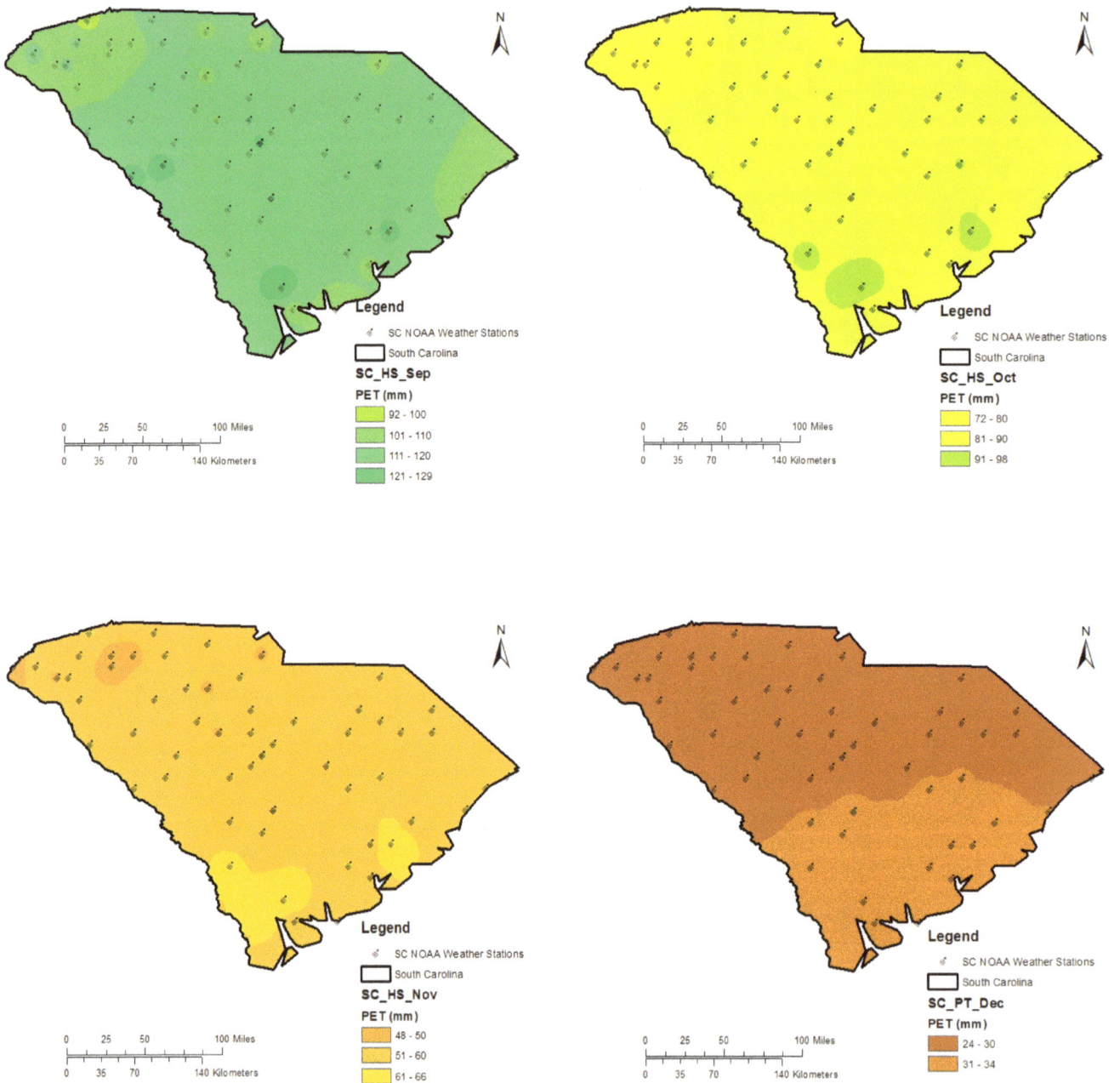

**Figure 6c.** Interpolated mean monthly H-S PET for September and October (top row) and November and December (bottom row). Data period of analysis is 1996-2015.

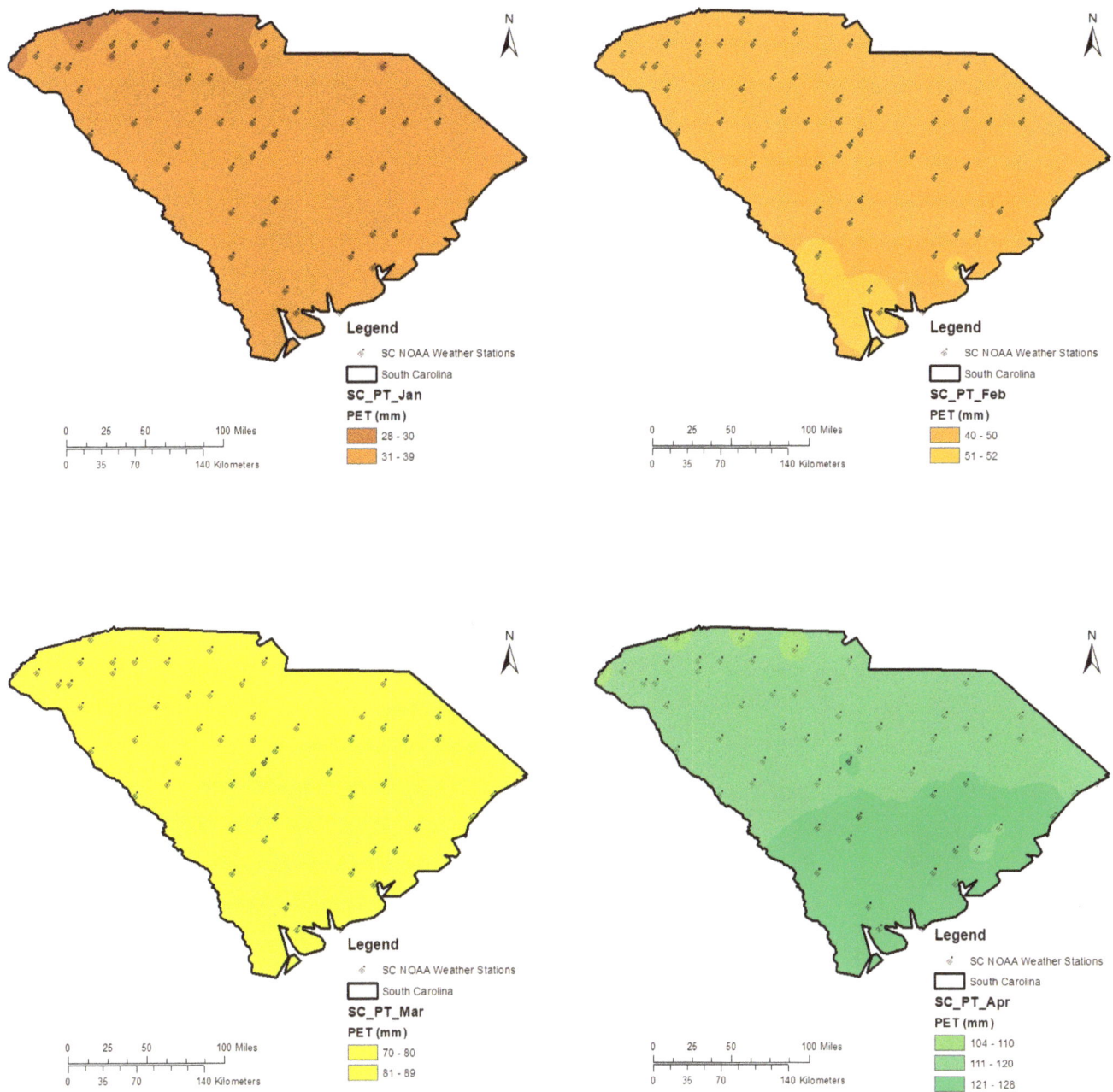

**Figure 7a.** Interpolated mean monthly P-T PET for January and February (top row) and March and April (bottom row). Data period of analysis is 1998-2015.

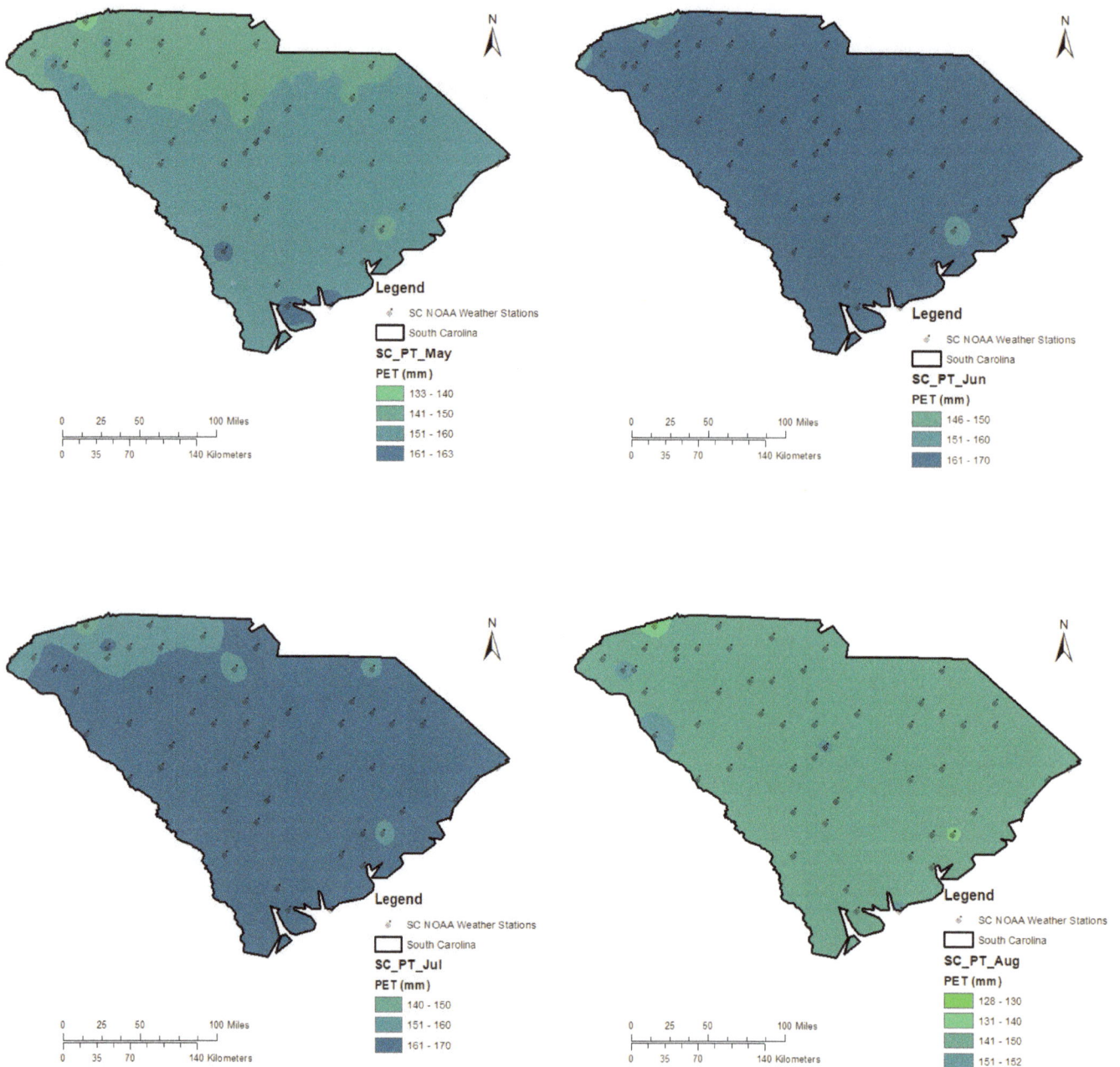

**Figure 7b.** Interpolated mean monthly P-T PET for May and June (top row) and July and August (bottom row). Data period of analysis is 1998-2015.

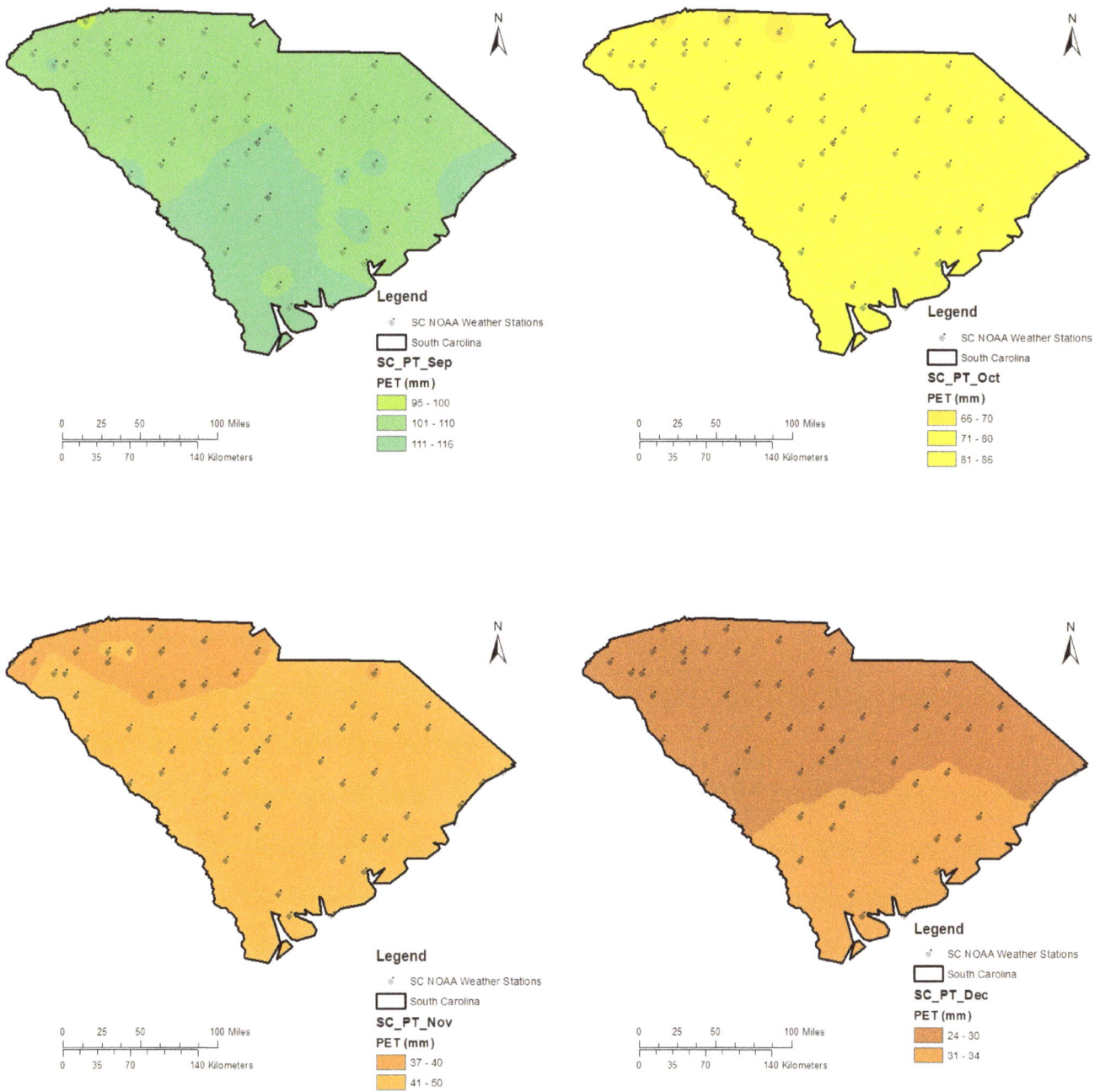

**Figure 7c.** Interpolated mean monthly P-T PET for September and October (top row) and November and December (bottom row). Data period of analysis is 1998-2015.

## SUMMARY AND CONCLUSIONS

Based on the long term (1998–2015) data, both the mean annual calculated H-S and P-T PET for Mountain region were significantly ($P < 0.05$) lower than for the Coastal and Piedmont regions. There was no significant difference in H-S PET between Coastal and Piedmont regions. The observed spatial annual mean PET trend, Coastal > Piedmont > Mountain, by each of the 3 methods (H-S, P-T, and P-M) using shorter term (2002–2009) data, was similar to measured weather variables (air temperature, solar radiation, and relative humidity) trend, with the highest and the lowest PET observed during summer and winter months, respectively. The mean annual P-M PET was found to be larger than both the H-S and P-T PET in both the Coastal and Piedmont regions based on the limited site-year data. Thus regional differences in weather variables and their influences found on estimated PET by 3 widely used methods will give water managers and policy-makers valuable information for water resource management and planning in South Carolina. However, based on the NSE and PBIAS evaluation statistics it was concluded that the adjusted H-S method performed better than the P-T when compared with the standardized P-M PET (REF-ET) as well as PE in this study, although a slope closer to unity and slightly higher $R^2$ was found for the P-T than for the H-S PET when compared with the PE. Limited stations with short-term record of complete dataset prevented us from concluding about the standard P-M PET method in this study. At the same time, considering the forest as dominant land use in South Carolina and changing climatic pattern in the southeast, energy-balance-based P-T method may ultimately be a choice for regional water management decisions including for OWE from lakes as the PET is strongly influenced by radiation also besides the air temperature used in the H-S method. However, more rigorous ground-truthing of publically available modeled solar radiation data used in this method is warranted as more data becomes available for its operational application. Furthermore, future studies should also test the reliability of these PET methods either by using simple water balance from gauged catchments or using hydrologic models. It is also recommended that the pan factors be derived using widely recommended empirical formulas involving climatic variables measured at the PE stations for assessing open water evaporation.

Results from this study can be used to support several components of the ongoing water planning efforts in South Carolina. For example, improved estimates of OWE on reservoirs can be incorporated into surface water modeling applications such as the simplified water allocation model, the model currently being used to help assess surface-water availability in the 8 major river basins in the state. Similarly, estimates of PET as reference ET can be used to estimate crop irrigation requirements for estimating future water demands and also possibly as inputs for the groundwater flow models being developed in the Coastal Plain to assess groundwater availability. These findings and the associated methods are easily transferrable to other states and regions that have similar needs and available data.

## ACKNOWLEDGEMENTS

This study was funded by the South Carolina Department of Natural Resources under an agreement with the College of Charleston. The authors acknowledge Andy Harrison, Hydrologic Technician at USDA Forest Service SEF and staff at Savannah River Site, South Carolina, for helping with data sharing and processing from those 2 weather stations. The authors also sincerely acknowledge Dr. Garry Grabow, Professor at North Carolina State University, Dr. Hamid Farahani at Natural Resources Conservation Service, and the anonymous reviewers for providing constructive suggestions to improve the manuscript quality.

## LITERATURE CITED

Allen RG. 1997. Self-calibrating method for estimating solar radiation from air temperature. ASCE J Hydrol Eng. 2(2):56–67.

Allen RG, Pereira LS, Raes D, Smith M. 1998. Crop evapotranspiration: guidelines for computing crop water requirements (Irrigation and Drainage Paper 56). Rome (Italy): U.N. Food and Agriculture Organization.

Allen RG, Pruitt WO, Wright JL, Howell TA, Ventura F, Snyder R, Itenfisu D, Steduto P, Berengena J, Yrisarry JB, et al. 2006. A recommendation on standardized surface resistance for hourly calculation of reference ETo by the FAO56 Penman-Monteith method. Agric Water Manag. 81:1–22.

Amatya, DM, Dai Z, Tian S, Sun G. 2016. Long-term PET and ET of two different forests on the Atlantic Coastal Plain. ET Special Collection, Trans. of the ASABE, 59(2): 647–660.

Amatya DM, Harrison CA. 2016. Grass and forest potential evapotranspiration comparison using five methods in the Atlantic Coastal Plain. J Hydrol Eng. 21(5) 05016007:1-13.

Amatya DM, Skaggs RW, Chescheir GM, Fernandez GP. 2000. Solar and net radiation for estimating potential evaporation from three vegetation canopies (Paper Number 002135). Paper presented at the ASAE International Meeting, Milwaukee, WI.

Amatya DM, Skaggs RW, Gregory JD. 1995. Comparison of methods for estimating REF-ET. J Irrig Drain Eng. 121(6):427–435.

Archibald JA, Walter MT. 2014. Do energy-based PET Models require more input data than temperature-based models? An evaluation at four humid flux net sites. J Am Water Resour Assoc. 50(2):497–508.

Barker RA, Pernik M. 1994. Regional hydrology and simulation of deep ground-water flow in the Southeastern Coastal Plain aquifer system in Mississippi, Alabama, Georgia, and South Carolina (Professional Paper 1410-C). Denver (CO): U.S. Geological Survey.

Barik M, Hogue TS, Franz K, Kinoshita AM. 2016. Assessing satellite and ground-based potential evapotranspiration for hydrologic applications in the Colorado River Basin. J Am Water Resour Assoc. 52(1):48–66.

Brauman KA, Freyberg DL, Daily GC. 2012. Potential evapotranspiration from forest and pasture in the tropics: a case study in Kona, Hawaii. J Hydrol. 440–441:52–61.

Cai J, Liu Y, Lei T, Pereira LS. 2007. Estimating reference evapotranspiration with the FAO Penman–Monteith equation using daily weather forecast messages. Agric For Meteorol. 145:22–35.

CDM Smith. (2016). Evaporation data and methodology: a technical memorandum from CDM Smith to SC Department of Natural Resources as a part of South Carolina Surface Water Quantity Modeling Project. Columbia (SC): South Carolina Department of Natural Resources and South Carolina Department of Health and Environmental Control.

Chattopadhyay N, Hulme M. 1997. Evaporation and potential evapotranspiration in India under conditions of recent and future climate change. Agric For Meteorol. 87:55–73.

Chen T, Ren L, Yuan F, Yang X, Jiang S, Tang T, Zhang L. 2017. Comparison of spatial interpolation schemes for rainfall data and application in hydrological modeling. Water. 9(5):342.

Dai Z, Trettin CC, Amatya DM. 2013. Effects of climate variability on forest hydrology and carbon sequestration on the Santee Experimental Forest in Coastal South Carolina (General Technical Report SRS-172). Ashville (NC): USDA Forest Service Southern Research Station.

FAO. (1990). "Report on the expert consultation on revision of FAO methodologies for crop water requirements." Land and Water Development Division, Food and Agriculture Organization of the United Nations, Rome, Italy.

Farnsworth RK, Thompson ES, Peck EL. 1982. Evaporation atlas for the contiguous 48 United States (Technical Report NWS 33). Washington (DC): National Oceanic and Atmospheric Administration.

Flint AL, Childs SW. 1991. Use of the Priestley–Taylor evaporation equation for soil water limited conditions in a small forest clearcut. Agric For Meteorol. 56:247–260.

Garcia MD, Raes D, Allen R, Herbas C. 2004. Dynamics of reference evapotranspiration in the Bolivian highlands (Altiplano). Agric For Meteorol. 125:67–82

Geraghty JJ, Miller DW, Van der Leeden F, Troise FL. 1973. Water Atlas of the United States. Port Washington (NY): Water Information Center.

Grismer ME, Orang M, Snyder R, Matyac R. 2002. Pan Evaporation to Reference Evapotranspiration Conversion Methods. J Irrig Drain Eng. 128(3):180–184.

Hamon WR. 1963. Computation of direct runoff amounts from storm rainfall. Int Assoc Sci Hydrol. 63:52–62.

Harder SV, Amatya DM, Callahan TJ, Trettin CC, Hakkila J. 2007. Hydrology and water budget for a forested Atlantic coastal plain watershed, South Carolina. J Am Water Resour Assoc. 43(3):563–575.

Hargreaves, G.H., and Z.A. Samani. 1982. Estimating potential evapotranspiration. J. Irrig. and Drain Engr., ASCE, 108(IR3):223-230.

Hargreaves, G.H., and Z.A.Samani. 1985. Reference crop evapotranspiration from temperature. Appl Eng Agric. 1:96–99.

Hember HA, Coops NC, Spittlehouse D. 2017. Spatial and temporal variability of potential evaporation across North American forests. Hydrol. 4(5):1–23.

Irmak S, Haman DZ, Jones JW. 2002. Evaluation of Class A pan coefficients for estimating reference evapotranspiration in humid location. J Irrig Drain Eng. 128(3):153–159.

Jensen ME, Allen RG (Eds). 2016. Evaporation, Evapotranspiration, and Irrigation Water Requirements. 2nd Edition, Manuals and Reports on Engineering Practice No. 70, New York, (NY): American Society of Civil Engineers.

Jensen ME, Burman RD, Allen RG. (Eds.). 1990. Evapotranspiration and irrigation water requirements (Manuals and Reports on Engineering Practices No. 70). New York (NY): American Society of Civil Engineers.

Lackstrom K, Carbone GJ, Tufford DL, Patel A. 2016. Climate change and water resources in the Carolinas: producing locally relevant information from global scenarios. J South Carolina Water Resour. 3(1):49–61.

Lang D, Zheng J, Shi J, Liao F, Ma S, Wang W, Chen X, Zhang M. 2017. A comparative study of potential evapotranspiration estimation by eight methods with FAO Penman–Monteith method in Western China. Water. 9:734. doi:10.3990/w9100734.

Lima JR, Antonino ACD, de Souza ES, Hammecker C, Montenegro SMGL, Lira CABdO. 2013. Calibration of Hargreaves–Samani equation for estimating reference evapotranspiration in sub-humid region of Brazil. J Water Resource Prot. 5:1–5.

Lopez-Moreno JI, Hess TM, White SM. 2009. Estimation of reference evapotranspiration in a mountainous Mediterranean site using the Penman–Monteith equation with limited meteorological data. Pirineos. 164:7–31.

Lu J, Sun G, McNulty SG, Amatya DM. 2003. Modeling actual evapotranspiration from forested watersheds across the Southeastern United States. J Am Water Resour Assoc. 39(4):887–896.

Lu J, Sun G, McNulty SG, Arnatya DM. 2005. A comparison of six potential evapotranspiration methods for regional use in the Southeastern United States. J Am Water Resour Assoc. 41(3):621–633.

Maidment DR. 1993. Handbook of hydrology. New York (NY): McGraw-Hill. Chapter 4, Evaporation.

Makkink GF. 1957. Testing the Penman formula by means of lysimeters. J Inst Water Eng. 11:277–288.

Marek G, Gowda P, Marek T, Auvermann B, Evett S, Colaizzi P, Brauer D. 2016. Estimating preseason irrigation losses by characterizing evaporation of effective precipitation under bare soil conditions using large weighing lysimeters. Agric Water Manag. 169:115–128.

McCuen RH. 1989. Hydrologic analysis and design. Upper Saddle River (NJ): Prentice Hall.

McKinney MS, Rosenberg NJ. 1993. Sensitivity of some potential evapotranspiration estimation methods to climate change. Agric For Meteorol. 64:81–110.

McMahon TA, Peel MC, Lowe L, Srikanthan R, T. R. McVicar TR. 2013. Estimating actual, potential, reference crop and pan evaporation using standard meteorological data: a pragmatic synthesis. Hydrol Earth Syst Sci. 17:1331–1363.

Mizzell H, Malsick M, Abramyan I. 2014. South Carolina's climate report card: understanding South Carolina's climate trends and variability. J South Carolina Water Resour. 1(1):4–9.

Monteith JL. 1965. Evaporation and environment. Symp Soc Exper Biol. 19:205–234.

Moriasi DN, Arnold JG, Van Liew MW, Bingner RL, Harmel RD, Veith TL. 2007. Model evaluation guidelines for systematic quantification of accuracy in watershed simulations. Trans. ASABE 50(3):885-900.

Natural Resources Conservation Service (NRCS). 2016. National engineering handbook. Columbia (SC): United States Department of Agriculture, Natural Resources Conservation Service. Part 652: Irrigation, Supplement– South Carolina irrigation guide.

Penman HL. 1948. Natural evaporation from open water, bare soil, and grass. Proc. Royal Soc. London A. 193(1032): 120-145. http://dx.doi.org/10.1098/rspa.1948.0037.

Phillips R, Saylor JR, Kaye NB, Gibert J. 2014. A comparison of remote sensing estimates of lake evaporation with pan evaporation measurements along the Savannah River Basin. Proceedings of the 2014 South Carolina Water Resources Conference, Columbia, SC.

Phillips RC, Saylor JR, Kaye NB, Gilbert JM. 2016. A multi-lake study of seasonal variation in lake surface evaporation using MODIS satellite-derived surface temperature. Limnology. 17:273–289. doi:10.1007/s10201-016-0481-z.

Priestley CHB, Taylor RJ. 1972. On the assessment of surface heat flux and evaporation using large-scale parameters. Mon Weather Rev. 100:81–92.

R Core Team (2017). R: A language and environment for statistical computing. R Foundation for Statistical Computing, Vienna, Austria. URL https://www.R-project.org/.

Rao LY, Sun G, Ford CR, Vose JM. 2011. Modeling potential evapotranspiration of two forested watersheds in the Southern Appalachians. Trans ASABE. 54(6):2067–2078.

Raziei T, Pereira LS. 2013. Spatial variability analysis of reference evapotranspiration in Iran utilizing fine resolution gridded datasets. Agric Water Manag. 126:104–118.

Roehl EA Jr, Conrads PA. 2015. Decision support system for optimally managing water resources to meet multiple objectives in the Savannah River Basin. J South Carolina Water Resour. 2(2):16–23, 2015.

Rosenberry DO, Winter TC, Buso DC, Likens GE. 2007. Comparison of 15 evaporation methods applied to a small mountain lake in the northeastern USA. J Hydrol. 340:149–166.

Shevenell L. 1996. Statewide potential evapotranspiration maps for Nevada (Report 48). Reno (NE): Nevada Bureau of Mines and Geology.

Shukla J, Mintz Y. 1982. Influence of land-surface evapotranspiration on the Earth's climate. Science. 215(4539):1498–1501.

Singh VP, editor. 2016. Handbook of applied hydrology. 2nd ed. New York (NY): McGraw Hill.

Ssegane H, Amatya DM, Muwamba A, Chescheir GM, Appelboom T, Tollner EW, Nettles JE, Youssef MA, Birgand F, Skaggs RW. 2017. Calibration of paired watersheds: utility of moving sums in presence of externalities. Hydrol Process. 31(20):3458–3471. doi:10.1002/hyp.11248.

Sumner D, Geurink J, Swancar A. 2017. Use of eddy-covariance methods to "calibrate" simple estimators of evapotranspiration (Paper Number 1700912). Presented at: 2017 American Society of Agricultural and Biological Engineers Annual International Meeting, Spokane, WA. DOI: 10.13031/aim.201700912.

Sun G, Alstad K, Chen J, Chen S, Ford CR., Lin G, Zhang Z. 2011. A general predictive model for estimating monthly ecosystem evapotranspiration. Ecohydrology. 4(2): 245–255. doi:10.1002/eco.194.

Tegos A, Malamos N, Koutsoyiannis D. 2015. A parsimonious regional parametric evapotranspiration model based on a simplification of the Penman–Monteith formula. J Hydrol. 524:708–717.

Thomas A. 2000. Spatial and temporal characteristics of potential evapotranspiration trends over China. Int J Climatol. 20:381–396.

Thornthwaite CW. 1948. An approach toward a rational classification of climate. Geograph Rev. 38(1):55–94.

Turc L. 1961. Estimation of irrigation water requirements, potential evapotranspiration: a simple climatic formula evolved up to date. Ann Agronomy. 12:13–49.

Wang W, Xing W, Shao Q. 2015. How large are uncertainties in future projection of reference evapotranspiration through different approaches? J Hydrol. 524: 696–700.

Winter TC, Rosenberry DO, Sturrock AM. 1995. Evaluation of 11 equations for determining evaporation for a small lake in the north central United States. Water Resour Res. 31(4):983–993.

Young CE. 1968. Water balance of a forested coastal plain watershed on the Santee Experimental Forest (Technical Report FS-SE-1603). Asheville (NC): USDA Forest Service.

Journal of South Carolina Water Resources, Volume 5, Issue 1, Pages 25–33, 2018

# Spatial Analysis of Hydrological Productivity in Fractured Bedrock Terrains of the Piedmont of Northwestern South Carolina

Brooks Bailey[1], Weston Dripps[1*], and Suresh Muthukrishnan[1]

AUTHORS: [1]Department of Earth and Environmental Sciences, Furman University, Greenville, SC 29613.
*weston.dripps@furman.edu

**Abstract.** Fractured bedrock aquifers are structurally complex groundwater systems. Groundwater flow is limited to secondary porosity features such as faults and fractures on account of the low primary porosity and permeability of the native bedrock. The hydrologic productivity of wells drilled within these systems is spatially and vertically variable because of limited interconnectivity among these features. The purpose of this study was to assess potential correlations between driller-estimated well yields and the mapped lithology and structural features of the fractured bedrock aquifers of the Piedmont of northwestern South Carolina. Groundwater well data (e.g., well depth, well yields, static water level) of 1,069 wells, geologic data (e.g., lithology, mapped structural features), and topographic data (e.g., surface elevation, slope) were integrated within a geographic information system database for a spatial analysis of well yield distribution. Wells drilled in alluvium had the highest median yield (15 gal/min), whereas those drilled in schist, amphibolite, and gneisses had lower median yields (9, 8.5, and 8 gal/min, respectively). Nonparametric statistical analyses indicated that no geologic or topographic variables considered were strongly or moderately correlated with reported well yields. Spearman's correlation coefficients for well depth (0.24), static water level (0.19), proximity to water bodies (–0.10), and proximity to lithologic contacts (–0.08) were statistically significant (at the 0.05 confidence level) but only weakly correlated with well yield. Topographic variables and proximity to mapped faults were not statistically significant. Wells drilled in alluvium had the highest yields due to the higher porosity and permeability compared to the bedrock. However, alluvium makes up less than 5% of the study area surface, and so opportunities to further tap this unit are limited and spatially constrained. The lower median yields of other lithologies are attributed to the lack of fracture development in amphibolite and the low degree of weathering within gneiss foliation planes. To maximize yields, wells should be drilled in alluvium close to water bodies and lithologic contacts where possible.

## INTRODUCTION

Understanding the hydrology of fractured rock terrains remains one of the most challenging and complex problems in water resources management and development. The challenges stem from the inherent structural complexities of aquifers in fractured crystalline bedrock (Moore et al., 2002). Bedrock is typically characterized by low matrix permeability and porosity with flow largely governed by secondary porosity features such as fractures and faults (Boutt et al., 2010). Water availability in fractured rock terrains is spatially and vertically variable and can range over several orders of magnitude among lithologies and over relatively short distances due to heterogeneous fracture distribution and variable degrees of interconnectivity between structural features (Shapiro et al., 1999). The Piedmont of northwestern South Carolina is a structurally complex, fractured igneous and metamorphic rock terrain with geology that has been surficially studied and mapped over the past 4 decades (Garihan, 2009; Garihan et al., 2005; Gellici, 1989; Griffin, 1974; Mitchell, 1995; Shapiro et al., 1999). During this same time period, there has been a proliferation of groundwater wells drilled in the regional fractured crystalline bedrock for domestic, agricultural, and municipal use (Gellici, 1989), but no previous attempts have been made to relate the observed structural features and lithological units to the subsurface hydrology of the region. With increasing demand for water resources, there is a greater need for understanding the relationship between the region's geology and hydrological productivity (Wachob et al., 2009). Identifying the structural and geological features associated with hydrologically productive areas will aid groundwater prospecting efforts and promote sustainable development of groundwater resources in fractured bedrock aquifers in this region and other fractured bedrock terrains.

Many previous researchers have attempted to identify and characterize hydrologically productive zones in fractured bedrock terrains in response to the increasing importance of fractured bedrock aquifers as a water source, particularly for rural populations in places where surface flow does not meet water demands (Henriksen, 1995; Mabee, 1999; Moore et al., 2002; Yin & Brook, 1992). These studies have shown that the factors controlling hydrologic productivity are numerous and vary by physiographic and geologic setting. Henriksen (1995) examined the relationship between topography and well yield in the crystalline bedrock of western Norway and found that boreholes drilled in flatlands and valley bottoms have significantly higher yields than those in fjords and valley slopes, presumably due to lower recharge rates associated with steeper topographic settings. Conversely, Yin and Brook (1992) observed no significant relationship between surface topography and hydrologic productivity in the Blue Ridge physiographic province of northeastern Georgia, but found that well depth and proximity to fracture traces had the greatest influence on well yield. Other studies (Edet et al., 1998; Magowe & Carr, 1999; Solomon & Quiel, 2006) have focused on the relationship between fracture traces and well yield. Water prospecting efforts in the crystalline bedrock settings of northeast, south, and central Africa have shown that high well yield is associated with proximity to fracture traces (Edet et al., 1998; Magowe & Carr, 1999; Solomon & Quiel, 2006). Moore et al. (2002) found a relationship between well yield and a number of factors including topographic slope and proximity to surface water bodies within the glaciated metamorphic terrain of New Hampshire. Mabee (1999) analyzed several variables in a study of hydrologic productivity in glaciated metamorphic bedrock of Maine and found a moderate positive relationship between bedrock type and structural position with well yield. Wells drilled in amphibolite near anticline limbs had the highest reported yields (Mabee, 1999). These studies, from various similar geological terrains and physiographic settings, highlight that the controlling factors of well yield appear to be variable and, in places, spatially dependent on a variety of structural, lithologic, and topographic features.

The fractured crystalline bedrock terrain of the Piedmont and Blue Ridge provinces of the Carolinas has been the focus of much previous geological and hydrological research. Daniel (1989) related well construction methods to well yield in western North Carolina and found that high yields were associated with deeper wells. There was a considerable scatter in yields for wells drilled in generalized geologic belts including the Blue Ridge, Chauga, Carolina Slate, and Charlotte belts. The Piedmont of South Carolina has been surficially studied and mapped on 1:24,000 topographic quadrangles. Mitchell (1995) conducted a survey of ground-water wells of Greenville County in conjunction with the South Carolina Department of Natural Resources, which provided a descriptive statistical, but not spatial, assessment of well productivity within the county. Snipes et al. (1983) examined the relationship between well yield and lithological unit with Abbeville County of northwestern South Carolina and found that regions with fractured rocks were more hydrologically productive than those without fractures. However, little work has been done to relate the mapped structural and geological features with the hydrology of the Piedmont, and to date, the controls of hydrologic productivity in the region remain largely unknown.

The purpose of this study was to assess potential correlations between driller estimated well yields and the mapped lithology and structural features of the fractured bedrock aquifers of the Piedmont of northwestern South Carolina. Results could be used to characterize hydrologically productive areas within the Piedmont of South Carolina based on their respective structural and geologic settings. Groundwater well data and geologic data were combined to explore potential controls of hydrologic productivity in the fractured bedrock of South Carolina and thereby improve our knowledge of complex fractured bedrock aquifers in other regions. Collectively, the work is intended to lead to better groundwater prospecting methodology and improved management strategies for these important water resources.

## STUDY AREA

The area investigated covers approximately 2,450 km$^2$ in the Piedmont region of northwestern South Carolina, defined by 15 U.S. Geological Survey 1:24,000-scale topographic quadrangles, and includes portions of northeastern Pickens County, northern Greenville County, and northwestern Spartanburg County (Figure 1). The study area spans from the gently rolling to hilly topography of the Piedmont physiographic province to the more rugged mountains and narrow valleys of the Blue Ridge physiographic province, with elevations ranging from 240 m to 900 m above sea level. The region has a humid subtropical climate, with warm to hot summers with daytime highs around 32° C and cold to mild winters with highs typically 5–10° C. Average annual precipitation varies across the Piedmont from 180 to 115 cm, decreasing from northwest to southeast largely due to the orographic effect of the Appalachian Mountain front (Cherry et al., 2001). Temporal precipitation distribution is relatively even across the year. During the summer, the main sources of rainfall are occasional tropical storms and regular afternoon thunderstorms produced by convective heating. During the winter, precipitation is primarily due to extratropical cyclones. Estimated annual recharge (precipitation-evapotranspiration) follows a similar spatial pattern to precipitation ranging from 100 cm to 40 cm from the mountains to the state's interior (Cherry et al., 2001).

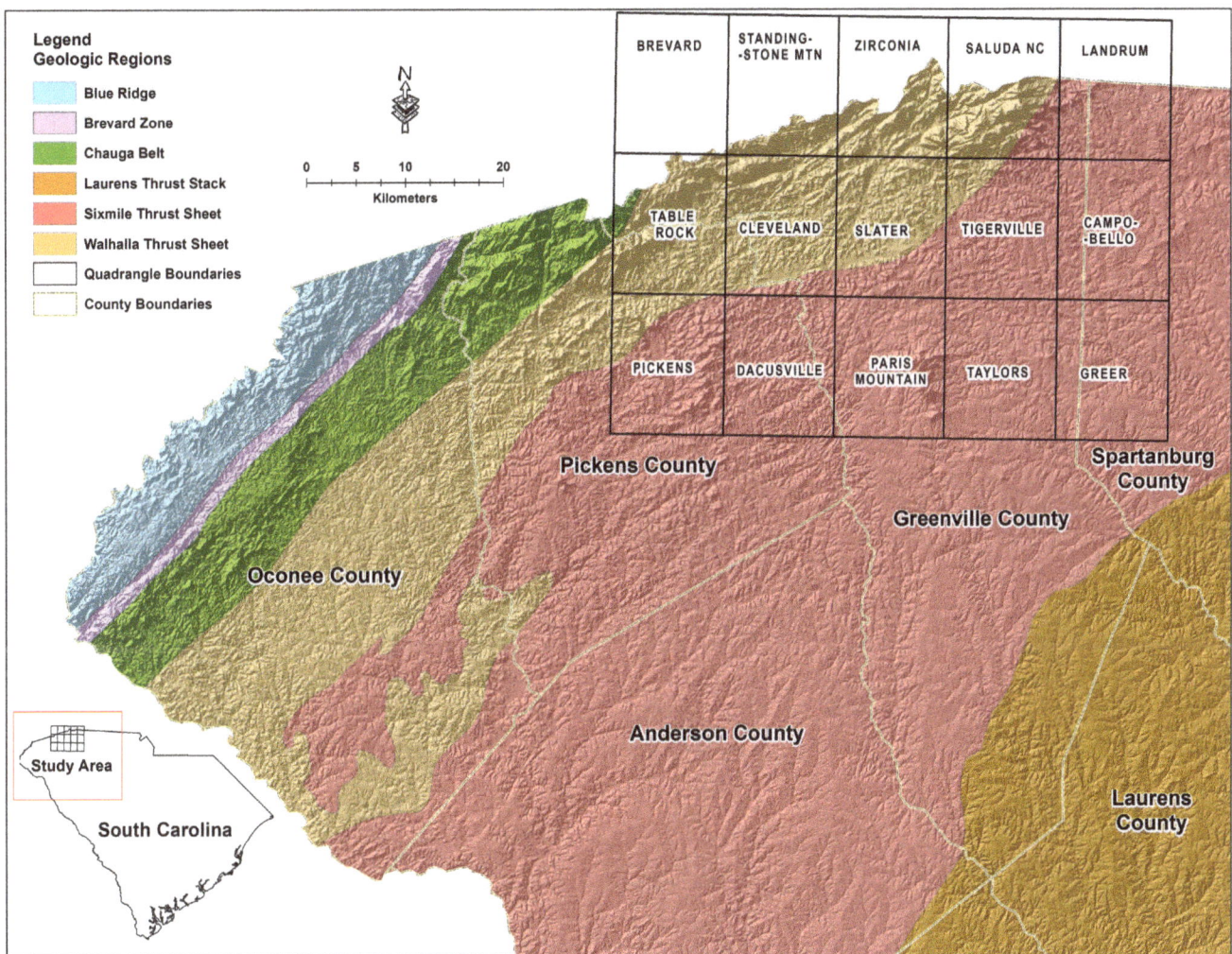

**Figure 1.** Study area composed of 15 topographic quadrangles in northwestern South Carolina. Major geologic regions are marked.

Geologically, the study area is located entirely within the inner belt of the Piedmont province (Willoughby et al., 2005). The prominent macroscopic structural features in the study area consist of the Six Mile and overlying Walhalla thrust sheet, a pair of westward-thrusting nappes trending northeast–southwest (Griffin, 1974). Other structural features include thrust faults, slip faults, synclines, anticlines, and diabase dikes. The majority of faults trend northeast–southwest, with dikes trending southwest–northeast. The underlying geology features a suite of metamorphic and igneous rocks with a metamorphic grade falling within the sillimanite zone of the amphibolite facies (Hatcher, 2002). The 4 main lithological map units include Poor Mountain Formation amphibolite (PMa; a well-foliated, slabby, fine- to medium-crystalline rock); Tallulah Falls Formation (TF; a mix of migmatitic and micaceous gneiss and schist); Table Rock gneiss (TRg; a biotite-rich quartzofeldspathic gneiss); and Quaternary alluvium (Qal; gravel, sand, silt, and clay deposits; Garihan et al., 2005; Figures 1 and 2).

The hydrology within the study area is controlled by a simplified, dual aquifer system consisting of regolith and fractured bedrock (Mitchell, 1995). The weathered regolith material overlying the bedrock, also known as saprolite, ranges in thickness from 3 m to 30 m (LeGrand, 1989). The saprolite zone is characterized by low permeability and high porosity and thus functions as a reservoir that feeds water into fractures within the underlying bedrock (LeGrand, 1989). Although the water storage capacity of fractured bedrock is low, water is capable of being transmitted along fractures and fracture intersections within the bedrock (Heath, 1980). The ability of these fractures to hold and transmit water diminishes with depth and tends to cease below about 30 m due to lithostatic pressure (Daniel, 1989).

## METHODS

This study integrates lithologic, structural, and hydrologic data in an attempt to better identify the controls on the complex fractured bedrock hydrology of the South Carolina Piedmont region. The data used came from a number of different sources and were compiled into ArcGIS software for spatial and statistical analyses.

Scale 1:24000 digital geological data were obtained for the study from the South Carolina Geologic Survey. The data were in digital Geographic Information System (GIS)-ready format for the 15 topographic quadrangles that cover the study area (Figure 1) and included many of the mapped surface features such as lithology, faults, tectonic folds, diabase dikes, brecciated rock zones, and water bodies. To perform the spatial analysis of well yield by lithology, the 43 reported mapped lithologic units from the original data were combined into 4 main lithologic groups including gneiss (TRg and TF subunits), amphibolite (PMa subunits), schist (TFs subunits and other micaceous schists), and alluvium (Qal) (Figure 2). Because structural features within the digital maps were organized as a mass of interconnected polyline features, individually mapped structural features were manually selected and extracted as separate, distinct shapefile feature classes to facilitate spatial analysis of well yield. The well data for the study area—which includes well depth, intended water use, estimated well yield, well log, drilling method, casing type, casing diameter, depth to bottom of casing, and static water level for each well—were obtained from South Carolina Department of Natural Resources.

Data are available, with varying degrees of precision for the 25,054 wells in 17 counties across the Piedmont of South Carolina. Only wells with localities that were known to the nearest second within the study area were selected. In all, this resulted in 1,069 wells that were imported into GIS and included in the study (Figure 2).

The state digital elevation model was obtained from the USGS at 1:250,000 with a cell size of 30 × 30 m. Topographic concavity and slope indices were extracted using the ArcGIS tools. Values for these data were extracted for each individual well locality to characterize the topographic setting for each well within the study area. The Near tool in ArcGIS was used to calculate straight-line distances (in meters) from wells to structural features within the quadrangle maps of the study area. Both water body density and fault density data were generated using the Line Density tool, which was used to calculate the density of faults and water body features within a circle with an area of 1 km² around each raster cell center. The goal in creating these thematic maps was to quantify the concentration of water bodies and fault zones, with the assumption that areas with a higher concentration of these features would be more hydrologically productive than those

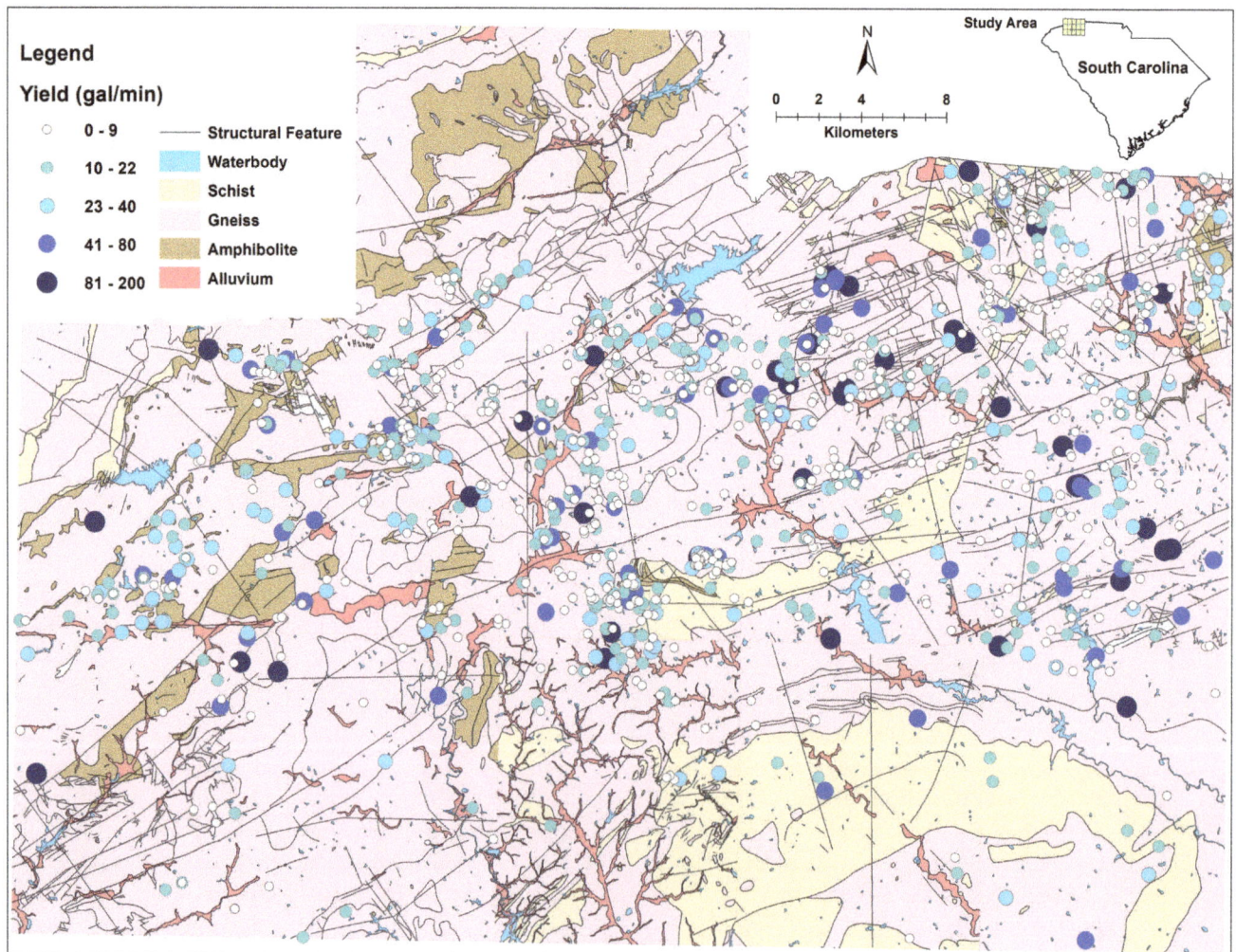

**Figure 2.** Generalized lithology, mapped structural features, and spatial distribution of well yields within study area.

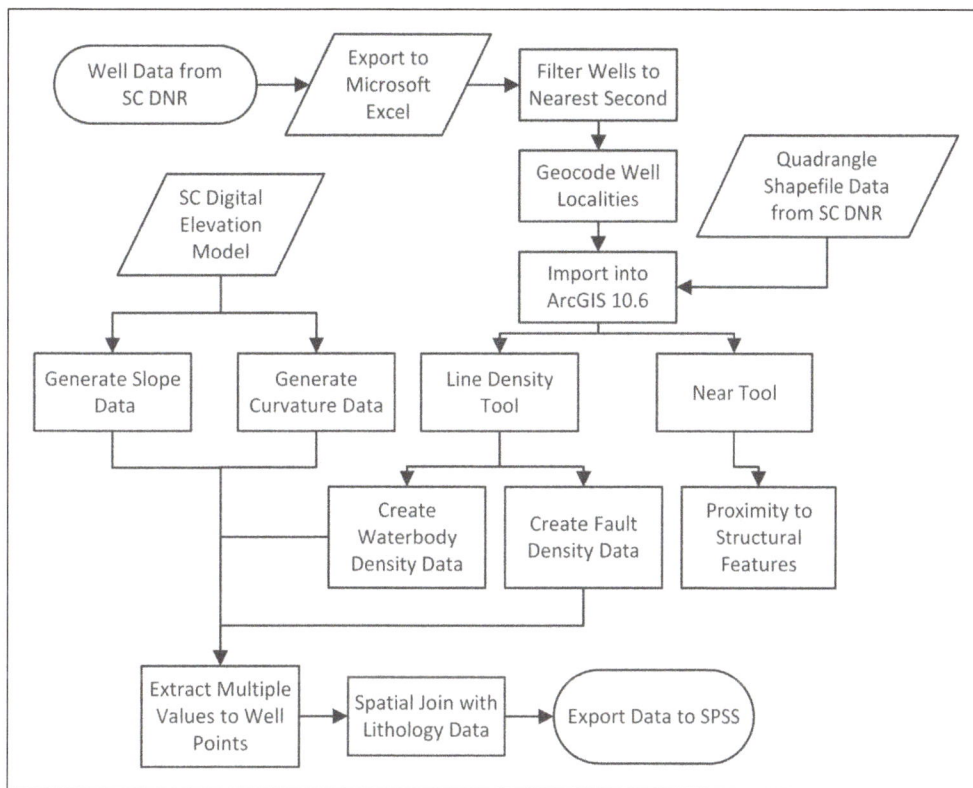

**Figure 3.** Workflow chart of procedures executed in ArcGIS.

with lower concentrations. The results of spatial analyses were then exported for statistical analysis (Figure 3).

## STATISTICAL TESTS

Statistical tests were performed on the well data using SPSS Version 20.0 (IBM Corp., 2011) to determine quantitative relationships between hydrological and geological factors. Driller-reported well yield (in gallons per minute) was used to represent each site's hydrological productivity. The Shapiro–Wilks test was used to assess the well yield data distribution. The null hypothesis of normal distribution was disproved ($P < 0.05$), indicating that the data were nonnormally distributed. The same procedure was repeated for natural-log-transformed data and yielded the same result. Because parametric statistical methods require normal distribution of data, the nonnormal, nonparametric statistical methods were used to check correlations within the positively skewed, nonnormal well yield data. Unlike parametric statistical methods, which test the differences in the means of data, nonparametric methods test the differences in the medians of data. Thus, nonparametric methods have less predictive power than parametric methods, but nonparametric methods still calculate direct correlations within data. Well yield was compared with both categorical and continuous data from the database, including well parameters (well depth, depth of casing, and static water level); well proximity to surface water bodies, topographic features (slope), lithologic features (contacts and diabase

dikes), structural features (synform axes, antiform axes, breccia zones, thrust faults, slip faults, and all faults); and density of water body and fault.

The Spearman's ρ rank correlation coefficient method was used to calculate correlations between well yield and other continuous variables within the database at the 0.05 significance level. To ensure correlation accuracy, continuous data including straight-line distances to mapped structural features were reclassified into groups of ordinal variables according to the methods of Moore et al. (2002). The Wilcoxon test was used to identify any significant differences in yield for wells grouped by simplified lithological units, following the methods of Henriksen (1995).

## RESULTS

### STATISTICAL AND SPATIAL ANALYSES OF WELL YIELD

Summary statistics for wells grouped according to lithology are shown in Tables 1 and 2. Wells drilled in alluvium have the highest mean and median yield. Median well yield is similar between schist, amphibolites, and undifferentiated gneisses. The number of wells drilled in gneisses is nearly 2 orders of magnitude higher than those in the other three lithologic units. Results of the Wilcoxon test show that differences in well yield between lithologically grouped samples are only statistically significant at the 0.05 confidence level between schist and alluvium.

**Table 1.** Summary statistics for wells grouped according to generalized surface lithology at drill site (Min. = minimum; Max. = maximum; total wells, $n = 1,069$; UG = undifferentiated gneisses; Am = amphibolite; S = schist; Al = alluvium).

|                | UG    | Am    | S     | Al    |
|----------------|-------|-------|-------|-------|
| Wells, $n$     | 956   | 36    | 58    | 19    |
| Min. yield, gpm| 0     | 2     | 0     | 1     |
| Max. yield, gpm| 200   | 50    | 50    | 45    |
| Median         | 8     | 8.5   | 9     | 15    |
| Mean, gpm      | 18.56 | 13.14 | 12.28 | 19.53 |
| SD, gpm        | 27.83 | 12.23 | 12.73 | 14.47 |

**Table 2.** Data matrix for generalized lithology (Y = significant and N = no significant difference between a pair of lithologically grouped samples according to the Wilcoxon test at the 0.05 confidence level; total wells, $n = 1,069$. UG = undifferentiated gneisses; Am = amphibolite; S = schist; Al = alluvium).

|    | UG | Am | S  | Al |
|----|----|----|----|----|
| UG | —  |    |    |    |
| Am | N  | —  |    |    |
| S  | N  | N  | —  |    |
| Al | N  | N  | Y  | —  |

**Table 3.** Results of Spearman's ρ correlation of well yield with continuous variables (*significant at the 0.05 confidence level).

| Variable | Spearman's ρ | P |
|----------|------|---|
| Well depth | 0.2433 | <0.0001* |
| Static water level | 0.1925 | <0.0001* |
| Slope | −0.0449 | 0.142 |
| Depth of casing | −0.0412 | 0.1785 |
| Curvature | −0.0345 | 0.2601 |
| Water body density | −0.0135 | 0.6595 |
| Fault density | −0.0086 | 0.7784 |
| Proximity to |  |  |
|   Surface water body | −0.1018 | 0.0009* |
|   Contacts | −0.0786 | 0.0101* |
|   Synforms | 0.0383 | 0.2106 |
|   Antiforms | 0.0383 | 0.2106 |
|   Breccia zone | −0.0281 | 0.359 |
|   Thrust faults | 0.037 | 0.2264 |
|   Slip faults | −0.0113 | 0.7121 |
|   All faults | 0.0169 | 0.5813 |
|   Diabase dikes | 0.0024 | 0.9385 |

Results of the Spearman's ρ correlation (Table 3) indicate that well depth, static water level, and proximity to surface water bodies and lithological contacts are the only statistically significant variables related to well yield. Despite their significance, these variables are only weakly correlated with changes in well yield. Of the variables considered, well depth is the highest correlated to well yield, with deeper wells associated with higher yields. Static water level elevation is the second highest correlated variable related to well yield, with higher static water level related to higher well yield. Well proximity to surface water bodies and lithological contacts are also weakly—very weakly—correlated to well yield. Wells closer to these features are associated with higher yields, as indicated by the negative Spearman ρ coefficients.

## DISCUSSION

The fractured bedrock aquifer of the Piedmont of northwestern South Carolina presents a hydrological challenge. Little is known regarding the structural, topographic, or lithological controls of hydrological productivity in the region. Whereas previous studies (Daniel, 1989; Gellici, 1989; Mitchell, 1995) provided descriptive statistical assessments of hydrological productivity, this study attempted to reveal the spatial relationships between the driller-estimated well yields and the mapped lithology and structural features of the region.

The high median yield of wells drilled in alluvium in this study was expected due to the high porosity and permeability of this rock type (Solomon & Quiel, 2006). However, alluvium makes up less than 5% of the study area surface (Figure 2), and so opportunities to further tap this unit will likely be limited and constrained spatially. The disparity between the observed mean (higher) and median (lower) yield values for undifferentiated gneisses is most likely due to the larger sample size ($n = 956$) and higher number of high-yield water supply wells drilled in this lithologic unit. Gneisses have shown sizable variability in well yield in several other studies due to composition, weathering, and expression of structural features (Chapman et al., 1999; Snipes et al., 1983; Solomon & Quiel, 2006). Solomon and Quiel identified foliation planes as permeability-enhancing structures within the gneisses of the central highlands of Eritrea. Chapman et al. (1999) also noted the enhanced weathering potential within the compositional layering of gneisses as a means of enhancing permeability and promoting greater groundwater flow. Biotite gneisses within their study area had the highest yields. Conversely, Snipes et al. (1983) found significantly lower yields within the granitic gneisses of South Carolina. Their reasoning behind the observed low productivities of this unit is attributed to its more massive composition, therefore making it more resistant to weathering. Based on the literature (Chapman et al., 1999; Mabee, 1999), the similarity in median and mean yield for wells drilled in amphibolites and schist (Table 1) was unexpected due to the greater degree of fracturing and greater fracture development typically associated with amphibolite units. Mabee (1999) observed greater fracture development and prevalence of steeply dipping, orthogonal fracture networks within the amphibolites of Maine, which he interpreted as the main reason for higher yields in wells drilled in this rock type than those drilled in schists. Chapman et al. (1999) observed prevalent jointing at depth within the amphibolites in upstate Georgia. According to their study, productive fracture zones exist at the intersection of low-angle compositional layering and joint surfaces, along which differential weathering enhances rock permeability. Both studies by Chapman et al. (1999) and Snipes et al. (1983) reported schists to be the least productive rock units within their respective studies. Chapman et al. (1999) identified low weathering potential and lack of jointing as possible reasons for the observed low yields within schist. The amphibolite rock units in this study have a similar fracture network to the schists (Figure 2) and do not appear to have as many well-developed fracture networks as those observed in other studies (Mabee, 1999; Moore et al., 2002; Solomon & Quiel, 2006). The observed homogeneity between yield values for wells drilled in undifferentiated gneisses, amphibolite, and schist contrasts with the results found by Daniel (1989) in his statistical study of well yield within the Piedmont of North Carolina. He found large variability between yields for wells drilled in various igneous and metamorphic rock units.

Well depth has the strongest relationship, albeit still a very low correlation, with well yield of the 16 variables of the study. This result is consistent with the findings of Moore et al. (2002) and Gellici (1989), who both found significant correlations between increasing well yield with greater well depth within the fractured bedrock aquifers of Maine and the Piedmont of South Carolina, respectively. Gellici noted that the hydrological productivity of deep wells is due in part to the prevalence of continuous, interconnected water-bearing fractures at depth.

In this study, static water level and proximity to surface water bodies were both significantly correlated with well yield. Presumably, these correlations are due to the greater prevalence of groundwater in wells with high static water levels and those proximal to water bodies. Mabee (1999) found no significant correlation between static water level and well yield, but did not offer any explanations for this result. Moore et al. (2002) found a significant inverse correlation between proximity to surface water bodies and well yield.

Proximity to lithological contacts had the weakest statistically significant correlation with well yield of all the variables. It is possible that the contact zones between lithological units within the study area are more transmissive due to the faults that sometimes occur along these contacts. Moore et al. (2002) noted that yield for wells drilled near fracture zones can change based on the lithological contacts crossed by the fracture zone. Moore et al. (2002) mentioned that wells situated near weathered, unconsolidated granitic rocks between unweathered plutonic rock units can have higher yields. Snipes et al. (1983) noted an increase in well yields along wells drilled near lithological contacts, especially contacts near the shattered country rock associated with zones of brecciation. Broken, fragmented brecciated zones should have higher secondary porosity and permeability compared to the bedrock, and as such, it should be more conducive to higher well yields.

Neither of the topographic variables, slope or curvature, was statistically correlated to well yield. Topographic variables were shown to be statistically significant in the fjords of Norway by Henriksen (1995), with higher yielding wells associated with valley bottoms and flatlands, presumably due to greater infiltration and recharge rates in these areas compared with slopes and peaks. Moore et al. (2002) also found a statistically significant relationship between slope and curvature, with high-angle slopes and concave-down regions associated with lower yields. However, Yin and Brook (1992) found no such relationship between well yield and topographic variables within the fractured bedrock terrain of the Georgia Piedmont, with topography explaining a mere 0.1% of variability in well yield. The topography of the Piedmont of Georgia is more similar to that of South Carolina than the topography found in other study areas, which seems to support the observed lack of correlation between topographic variables and well yield within the Inner Piedmont of South Carolina.

None of the structural features displayed a statistically significant correlation with well yield. Proximity to synforms and antiforms were not correlated with well yield. Mabee (1999) showed that wells located close to fold limbs generally had higher yields. However, he also mentioned that geological unit may have a greater control on well yield than proximity to fold limbs because wells near fold limbs within schist had lower yields than those located in the same structural position within amphibolites. Snipes et al. (1983) noted that structural features including synform axes were linked with increased well yield and explained that these structural features are commonly located along ridges, where the steep dips of the compositional bedding planes facilitate water movement, thus enhancing well productivity. In their study of the regionally folded and deformed fractured bedrock aquifer near the Lawrenceville, Georgia, area, Chapman et al. (1999) found a trend of increasing hydrological productivity for wells situated proximally to antiform axes. They proposed that the regional tectonic stresses that induced folding led to vertical joint development. These highly productive vertical joints are commonly located on or near the hinges of antiformal folds. It is surprising, then, that this study shows no relationship between well yield and proximity to fold features, given the observed positive trends found in other studies conducted in similar geologic regions. Likewise, proximity to zones of microbreccia was not correlated with well yield. This unconsolidated mass of coarse, angular rocks in a relatively finer grained matrix forms as a result of brittle deformation within a shear zone (Garihan, 2009). It is possible that the fractures associated with this shear zone do not reach the saprolitic regolith and are thus not transmissive. Proximity to diabase dikes was also not correlated with well yield. Although this rock type is characterized by low permeability and is often an aquitard, Chapman et al. (1999) noted that differential weathering between diabase dikes and the surrounding country rock can often result in the formation of preferential flow paths within the bedrock. It seems that if differential weathering has occurred in a similar manner within the study area for this study, then it has not produced any such flow paths within the bedrock. It is surprising that proximity to faults (thrust faults, synform, and antiform) and fault density were not correlated with well yield. Many previous studies (Edet et al., 1998; Mabee, 1999; Magowe & Carr, 1999; Moore et al., 2002; Solomon & Quiel, 2006) have consistently demonstrated that wells closer to faults and fracture zones are characterized by higher yields. There are several potential explanations for the lack of association in this study. It is possible that the fracture zones within the aquifer terminate prior to reaching the saprolite aquifer and thus do not transmit water (Mitchell, 1995). Another potential explanation for this trend is that the intense regional compression during emplacement of the nappes resulted in fracture zones with low degrees of interconnectivity and correspondingly low transmissivity.

There are several sources of error that may have influenced the results of this study. Driller-reported well yield is typically estimated on sight by drillers and is thus not always accurate or reliable. These estimates are made by forcing formation water out of the borehole via air pressure and then measuring the subsequent flow of this fluid over a short time interval. Pumping tests provide more accurate, long-term estimates of well yield, but driller-reported yield is favored by drilling companies in the interest of saving time and money (Mitchell, 1995). Thus, well yield can only be treated as a semiquantitative variable, which may distort the results of this study. Another potential source of error pointed out by other authors (Gellici, 1989; Mitchell, 1995; Moore et al., 2002) is that domestic wells are drilled based on the location of the owner's property and the economic constraints of the owner. Wells are not always drilled with the intention of achieving maximum yield; they are often drilled based on the needs of the owner, with the exception of water supply wells that often have higher yields because they are built to achieve maximum groundwater productivity. Because the majority of wells in this study were drilled in undifferentiated gneisses, sample size may be a source of error. Finally, error may also be present in using surface lithology as an indicator of well yield, because some of the deeper wells may tap rock units that are not expressed at the surface. This is a very likely source of error given that Chapman et al. (1999) observed vertical changes in lithology with depth due to intense folding and faulting in regionally deformed crystalline rock terrains.

## CONCLUSIONS

Although the correlations were weak ($|r| < 0.25$), the primary factors impacting hydrological productivity in the Piedmont of South Carolina based on driller-estimated well yields are well depth, static water level, and proximity to surface water bodies and lithological contacts. Similar to Yin and Brook's (1992) study, this article suggests that topography is not a driver of hydrological productivity in the Piedmont of South Carolina. Groundwater prospectors in this and similar regions should target alluvium units proximal to surface water bodies to maximize yields. Within gneisses, prospectors should target the transmissive fractures that seem to exist at depth. This article provides further evidence that fractured bedrock aquifers are among the most difficult water resources to characterize. As Mabee (1999) and Moore et al. (2002) pointed out, there is no universal driver for hydrological productivity in fractured bedrock terrains, and more research and field work are needed to enhance our understanding of these important groundwater resources.

## LITERATURE CITED

Bailey B, Dripps W, Muthukrishnan S. 2013. Spatial analysis of well yields in fractured bedrock terrains of the Piedmont of Northwestern South Carolina. Poster session presented at: Geological Society of America—Southeastern Section 62nd Annual Meeting; San Juan, Puerto Rico.

Boutt DF, Diggins P, Mabee S. 2010. A field study (Massachusetts, USA) of the factors controlling the depth of groundwater flow systems in crystalline fractured-rock terrain. Hydrogeol J. 18:1839–1854.

Chapman MJ, Crawford TJ, Tharpe WT. 1999. Geology and ground-water resources of the Lawrenceville area, Georgia (Water-Resources Investigations Report 98-4233). Atlanta (GA): U.S. Geological Survey.

Cherry R, Badr A, Wachob A. 2001. General hydrology of South Carolina. Columbia (SC): South Carolina Department of Natural Resources. http://www.dnr.sc.gov/water/hydro/HydroPubs/pdf/Map%202%20letter%20size.pdf.

Daniel CC. 1989. Statistical analysis relating well yield to construction practices and siting of wells in the Piedmont and Blue Ridge provinces of North Carolina (Water-Supply Paper 2341). Columbia (SC): South Carolina Department of Natural Resources.

Edet AE, Okereke CS, Teme SC, Esu EO. 1998. Application of remote-sensing to groundwater exploration: a case study of the Cross River State, southeastern Nigeria: Hydrogeol J. 6:304–404.

Garihan JM, Kalbas JL, Clendenin CW Jr. 2005. Geologic map of the Slater 7.5-minute quadrange, Greenville, County, South Carolina (South Carolina Geological Survey GQM-34). Columbia (SC): South Carolina Department of Natural Resources.

Garihan JM. 2009. Geologic map of the Landrum quadrangle, Greenville and Spartanburg Counties, South Carolina and Polk County, North Carolina (South Carolina Geological Survey GQM-44). Columbia (SC): South Carolina Department of Natural Resources.

Gellici JA. 1989. Borehole geophysics in the Piedmont of South Carolina. In: Daniel CC III, White RK, Stone PA, editors. Ground water in the Piedmont: proceedings of a conference on ground water in the Piedmont of the Eastern United States. Clemson (SC): Clemson University. p. 510–525.

Griffin VS Jr. 1974. Analysis of the Piedmont in northwest South Carolina. GSA Bull. 85(7):1123–1138.

Hatcher RD Jr. 2002. An Inner Piedmont primer. In: Hatcher RD Jr, Bream BR, editors. Inner Piedmont tectonics focused mostly on detailed studies in the South Mountains and the southern Brushy Mountains, North Carolina: Carolina Geological Society guidebook. Raleigh (NC): North Carolina Geological Survey. p. 1–18.

Heath RC. 1980. Basic elements of ground-water hydrology with reference to conditions in North Carolina (Water-Resources Investigations Open-File Report 80-44). Denver (CO): U.S. Geological Survey.

Henriksen H. 1995. Relation between topography and well yield in boreholes in crystalline rocks, Sogn og Fjordane, Norway. Groundw. 33(4):635–643.

IBM Corp. 2011. SPSS Statistics for Windows, version 20.0. Armonk (NY): Author.

LeGrand HE. 1989. A conceptual model of ground water settings in the Piedmont region. In: Daniel CC III, White RK, Stone PA, editors. Ground water in the Piedmont: proceedings of a conference on ground water in the Piedmont of the Eastern United States. Clemson (SC): Clemson University. p. 317–327.

Mabee SB. 1999. Factors influencing well productivity in glaciated metamorphic rocks. Groundw. 37(1):88–97.

Magowe M, Carr JR. 1999. Relationship between lineaments and ground water occurrence in western Botswana. Groundw. 37(2):282–286.

Mitchell HL. 1995. Geology, ground water, and wells of Greenville County, South Carolina (Hydrology–Water Resources Report 8). Columbia (SC): South Carolina Department of Natural Resources.

Moore RM, Schwarz GE, Clark SF Jr, Walsh GJ, Degnan JR. 2002. Factors related to well yield in the fractured-bedrock aquifer of New Hampshire (Professional Paper 1660). Denver (CO): U.S. Geological Survey.

Shapiro AM, Hsieh PA, Haeni GP. 1999. Integrating multidisciplinary investigations in the characterization of fractured rock. In: Morganwalp DW, Buxton HT, editors. U.S. Geological Survey Toxic Substances Hydrology Program: proceedings of the technical meeting–volume 3 of 3: subsurface contamination from point sources (Water Resources Investigations Report 99-4018C). Denver (CO): U.S. Geological Survey. p. 660–680.

Snipes DS, Padgett GG, Hughes WB, Springston GE. 1983. Ground water quantity and quality in fracture zones in Abbeville County, South Carolina (Technical Report no. 102). Clemson (SC): Clemson University Water Resources Research Institute.

Solomon S, Quiel F. 2006. Groundwater study using remote sensing and geographic information systems (GIS) in the central highlands of Eritrea. Hydrogeol J. 14:729–741. doi:10.1007/s10040-005-0477-y.

Wachob A, Park AD, Newcome R Jr. 2009. South Carolina state water assessment. 2nd ed. Columbia (SC): South Carolina Department of Natural Resources.

Willoughby RH, Howard CS, Nystrom PG. 2005. Generalized geologic map of South Carolina. http://www.dnr.sc.gov/geology/images/GGMS-1%20Poster%20x1_2011.pdf.

Yin Z-Y, Brook GA. 1992. The topographic approach to locating high-yield wells in crystalline rocks: Does it work? Groundw. 30(1):96–102.

Journal of South Carolina Water Resources, Volume 5, Issue 1, Pages 35–44, 2018

# Water Users' Perspectives:
# Summary of Withdrawal Survey Responses and Commentary

C. Alex Pellett[1] and Thomas Walker III[2]

AUTHORS: [1]Hydrologist, South Carolina Department of Natural Resources, 311 Natural Resources Drive, Clemson, SC 29631. [2]Postdoctoral Fellow, South Carolina Water Resources Center, 509 Westinghouse Road, Clemson University, Pendleton, SC 29670.

**Abstract.** The state of South Carolina is currently in a multiyear process of updating the State Water Plan, and water demand projections are an important component of that work. Predictions of water demand are inherently uncertain, but perhaps they can benefit from input by a diverse and robust sample of water users. A brief survey regarding water use was distributed to 780 permitted and registered water users in the state, including all water suppliers, industries, and irrigators withdrawing more than 3 million gallons in a month or more than 100,000 gallons in a day. There are 316 responses to 10 quantitative survey items that are summarized, presented, and discussed. Results indicate that most respondents plan to maintain their current levels of water use, consider their withdrawal reports to be accurate within 10%, and believe their current water supplies to be critical to their enterprise. A qualitative review of comments noted on survey responses includes a variety of potential drivers of water demand. The results motivate a discussion of recommendations for future research.

## INTRODUCTION

The purpose of this article is to describe trends that could be relevant for projecting future water demand. The costs of collecting detailed information on water use across the state can be substantial, and the time required to do so could render some information obsolete by the time the data collection process is considered complete. In this context, a short survey was devised and disseminated among permitted and registered water users in the state as a low-cost and efficient method to gain insight into current and future water demand. Quantitative and qualitative survey responses are summarized by use category, indicating trends in water demand, withdrawal reporting, and potential factors affecting future water use. Understanding current water use trends will inform estimation of future water demands, a key part of planning for water availability.

The specific objectives of this work are to (a) determine how water users' plans will impact their future water use, (b) investigate the accuracy with which water withdrawal data is reported, (c) assess the importance of current water supplies to water users' enterprises, and (d) compile a list of potential factors which could affect future water use in South Carolina.

In this report, water use is meant to include the withdrawal of fresh water from the environment and subsequent distribution of the water according to the socioeconomic motivations of humanity. In-stream uses such as hydropower and fishery habitat, though important, are not considered within the scope of this report.

### BACKGROUND

Mail and phone surveys have long been used to collect water use information (Holland, 1992). Although online water use reporting tools have also been used in some cases, mailed or downloaded forms and mail surveys continue to be available for water use reporting (Texas Water Development Board, 2017; South Carolina Department of Health and Environmental Control [SCDHEC], 2012).

As water planning in South Carolina has proceeded, many stakeholders have provided information regarding their use of and appreciation for the state's water resources. Water users who withdraw ≥100,000 gallons of water are required to obtain a permit or registration from SCDHEC. The mandatory permitting requirement came into effect July 1, 1983. Harrigan (1985) sought out reports of water volume and achieved an overall response rate of 67% after repeated mailings (Table 1).

The goal of the 1985 survey was to collect water usage information from all users believed to have a maximum single-day water usage ≥100,000 gallons. Power plants had a 96% response rate in the first mailing, and they are by far the largest water users in terms of volume. Excluding power

**Table 1.** 1983 Voluntary survey response rate of water users in South Carolina (Harrigan, 1985).

| Category | Sent | Responses | | Overall Rate |
| | | Mailing | | |
| | | First | Second | |
| --- | --- | --- | --- | --- |
| Public Supply | 268 | 101 | 81 | 68% |
| Industry | 432 | 346 | 55 | 93% |
| Agriculture | 681 | 269 | 126 | 58% |
| Power | 52 | 50 | 2 | 100% |
| Golf Courses | 180 | — | 51 | 28% |

plants, the 716 respondents to the first 1,381 surveys (golf courses were not included in the first mailing) represented about 75% of the remaining withdrawal water usage (Harrigan, 1985).

The second mailing of forms was put together with cover letters customized to different groups of nonrespondents based on likely reasons for not responding. Water use reporting provided official documentation of water use that could be used in case of conflicting demands. The South Carolina Water Resources Commission would use this information to identify water-deficient areas. Roughly a third of users who were obligated to report had already done so elsewhere, and duplicate reports were not required (Harrigan, 1985).

Following recommendations in water planning documents (Castro & Hu, 1997), reporting monthly water withdrawal volume became a mandatory annual ritual that now achieves a >99% overall response rate every year.

When the survey for the present study was distributed in 2017, there were 780 permitted and registered water users in the state. This population varies from farmers irrigating <100 acres to large power-generation projects including combustion and nuclear-powered thermoelectric generators operating in tandem with hydroelectric generators in multiple reservoirs. These various users are united in their reliance on a sufficient quantity of water to sustain their enterprise and dependent populations. The water users' contribution of time and effort in monitoring and reporting their monthly water withdrawal has enabled the compilation of a long-term dataset which can provide information on historic conditions and insight into current water use patterns. But to forecast future trends in water use, greater perspective is needed to provide context. Current water use patterns are less relevant if practices are expected to change in the future. If water withdrawers in South Carolina respond well to voluntary surveys, as demonstrated by Harrigan in 1985, then a similar survey could provide information with which to guide efforts to project future water demand.

## METHODS

The survey was composed of 20 items, including 3 used for identification purposes and 1 for follow-up email correspondence. The survey forms were attached to a cover letter describing the optional survey and with instructions on how to complete it (see Appendix for survey form and cover letter). A uniform survey and cover letter were prepared for all water users, with some specific instructions for irrigators and water suppliers.

The survey tool used to gather the quantitative and qualitative data for this research used a mailing survey, which was desirable for this research for several reasons. One benefit is the ease of distribution using a mailer that was already being sent to all registered or permitted users. Another benefit is that it was less duplicative and likely increased response rates. Additionally, some water users do not access the Internet, so using a digital-only survey would be exclusive. Digital survey tools, although not as desirable for this round of surveys, might be developed in subsequent years along with paper copies for mailing surveys.

The survey and attached cover letter were mailed by SCDHEC with the annual water withdrawal reporting forms to all registered and permitted users in the state. Envelopes were mailed the first week of December 2017, and follow-up envelopes were sent mid-February to permitted and registered water users who had not responded to the annual water withdrawal reporting forms.

## RESULTS

Results of this survey are presented corresponding to each of the 4 specific goals of the project. Results are divided between quantitative and qualitative summaries of survey responses and are presented for each category of water use. Response rates vary between survey items; not all respondents replied to all survey items, and some respondents marked multiple answers to some survey items.

### QUANTITATIVE SUMMARY

Table 2 summarizes the number of responses from each category of water withdrawers. Table 3 summarizes respondents' plans regarding the volume of water withdrawn at their enterprise over the next 5 years. Mining operations were most likely to expect an increase in their water withdrawals over the next 5 years. Water suppliers were the next most likely to expect an increase in their withdrawals. The majority of golf courses are expected to maintain current levels of withdrawal. Most agricultural and industrial withdrawals are also expected to maintain current volumes, but many expect to increase, and fewer expect to decrease. Table 4 summarizes responses on a similar topic, this time regarding the source of water over the next 5 years.

**Table 2.** 2017 Voluntary survey response rate of water users in South Carolina.

| Category | Sent | Responses | Rate |
|---|---|---|---|
| Agriculture | 327 | 159 | 49% |
| Golf Course | 157 | 59 | 38% |
| Industry | 83 | 33 | 40% |
| Mining | 14 | 7 | 50% |
| Thermoelectric | 17 | 17 | 100% |
| Water Supply | 190 | 58 | 31% |

**Table 3.** Water users' 5-year plans regarding withdrawal volume.

| Category | Increase | Maintain | Decrease | Don't Know |
|---|---|---|---|---|
| Agriculture | 47 | 72 | 5 | 34 |
| Golf Course | 5 | 40 | 6 | 11 |
| Industry | 6 | 18 | 4 | 5 |
| Mining | 6 | 1 | 0 | 0 |
| Water Supply | 33 | 15 | 1 | 10 |

**Table 4.** Water users' 5-year plans regarding water source.

| Category | More Surface | More Ground | More Purchased | Maintain | Don't Know |
|---|---|---|---|---|---|
| Agriculture | 8 | 42 | 1 | 81 | 31 |
| Golf Course | 3 | 3 | 0 | 46 | 7 |
| Industry | 4 | 2 | 2 | 21 | 6 |
| Mining | 2 | 4 | 0 | 1 | 0 |
| Water Supply | 11 | 26 | 3 | 12 | 7 |

**Table 5.** Number of respondents using different calculation methods by category of water use.

| Category | Flow Meter | Pumping Time | Pumping Energy | Estimation/ Reckoning |
|---|---|---|---|---|
| Agriculture | 21 | 109 | 10 | 20 |
| Golf Course | 44 | 18 | 1 | 0 |
| Industry | 25 | 8 | 1 | 2 |
| Mining | 0 | 7 | 0 | 0 |
| Water Supply | 56 | 5 | 0 | 0 |

**Table 6.** Responses estimating the precision of reported withdrawals by the various methods used to calculate withdrawal volume.

| Estimated Precision | Flow Meter | Pumping Time | Pumping Energy | Estimation/ Reckoning |
|---|---|---|---|---|
| Exact | 67 | 17 | 1 | 2 |
| ≤10% | 72 | 90 | 4 | 12 |
| ≤20% | 3 | 36 | 6 | 2 |
| >20% | 1 | 4 | 1 | 6 |

Figure 1 indicates that the perceived precision of water use reports varies among users and between different categories of use. As shown in Table 5, respondents in different categories tend to rely on different methods with which to calculate their monthly withdrawal volume. Table 6 compares the methods of calculation with the perceived precision of the reported volume; respondents tended to expressed high confidence in reported volume estimates derived from flow meters.

Figure 2 shows that most of the responding water users' enterprises are critically reliant on their current water supplies—their enterprises would cease without adequate water availability. Nevertheless, most respondents are not very concerned about future water availability for their enterprise (Figure 3). Notably, 100% of respondents representing mining enterprises described their operations

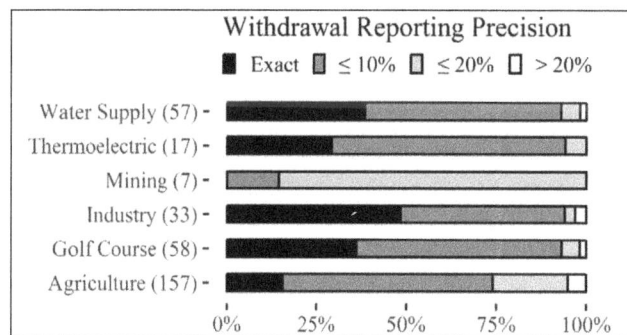

**Figure 1.** Water users' estimated reporting precision by category of use.

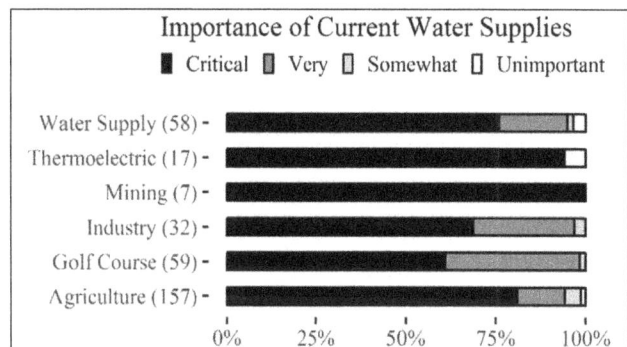

**Figure 2.** The importance of current water supplies to the continuation of water using enterprises summarized by category.

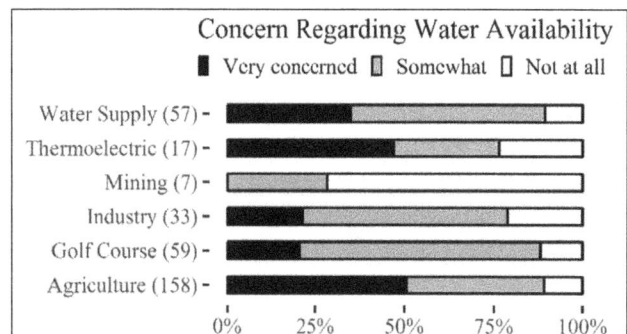

**Figure 3.** Water users' level of concern regarding the availability of their water supplies in the future, summarized by category.

as critically reliant on their water supplies, but none were very concerned regarding future availability, and most were not at all concerned about a shortage.

Finally, Table 7 summarizes some additional information that came from the survey responses. Some users are aware of existing studies that project water use at their enterprise. Respondents answered whether they purchase water in addition to their withdrawal volume and whether they sell water wholesale to water distributors.

**Table 7.** Additional information in the survey responses.

| Category | Existing Studies | Purchase | Wholesale | Reuse |
|---|---|---|---|---|
| Agriculture | 10 | 12 | 0 | 19 |
| Golf Course | 7 | 3 | 0 | 12 |
| Industry | 5 | 24 | 1 | 21 |
| Mining | 0 | 0 | 0 | 6 |
| Water Supply | 9 | 12 | 20 | 7 |

Industries, thermoelectric power plants, golf courses, water suppliers, and agricultural irrigators indicated that some amount of reclaimed or recycled water was used in their enterprises. The survey item did not distinguish between on-site reuse of water within an enterprise and reclamation of water discharged from another enterprise, although comments indicate that both kinds of water reuse occur in South Carolina.

## QUALITATIVE SUMMARY

The survey included 2 open-ended questions designed to elicit responses listing potential factors or trends that could impact water use in the future. Factors relating to the enterprises' economic markets and relevant developments in technology were sought. About half of respondents left these items blank or responded with something to the effect of "none," "do not know," or "not applicable." The remaining responses are summarized here for each category of water use.

**Agriculture.** Of the agricultural water users who responded to this question, the most common responses regarded commodity price fluctuations. If corn prices rise, then more corn will be planted, and corn requires relatively higher levels of irrigation. Some comments indicated that when commodity prices are low, irrigation becomes more important; other comments seemed to contradict that view—respondents indicated that commodity prices could be so low that the costs of operating irrigation equipment might not be offset by the increased yield. Corn was the most commonly used example in the responses, but other crops mentioned include pine, hay, sorghum, and sod. Among sod producers, increasing consumer demand for drought-tolerant varieties was a market factor that is expected to decrease their water

demand. However, it was also noted that an increase in housing development would increase total demand for sod (and other landscaping plants), which could increase the water demand of producers.

Many respondents noted, in response to these questions or in comments in the margins of other questions, that their water use is dependent on weather. The relationship between irrigation and weather is well established and could even be considered an *a priori* assumption in water demand forecasting. Other notable responses included an increasing demand for locally grown produce, developments in the North American Free Trade Agreement, and solar panel lease agreements.

Technological advances that commenters believed could impact their water use include new drought-resistant crop varieties, variable-rate irrigation allowing different irrigation depths on different soil types within the same field, moisture probes or sensors, drip irrigation, row covers, no-till or strip-till combined with cover crops, drip nozzles, unmanned aerial vehicles (i.e., drones) providing an overhead perspective on crop condition, and GPS-enabled irrigation equipment.

**Golf courses.** Among respondents representing golf course irrigation, several comments regarded changing perspectives on how golf courses should be maintained: Respondents noted the desire among some players for "firm and fast conditions," as well as a growing acceptance of "brown is the new green" and allowing for more natural vegetation in out-of-play areas of the course. These 3 trends in the golf course market allow for decreased irrigation; they also tend to make play more challenging.

Technological factors that were noted include the development and application of enhanced "wetting agents"; more drought-tolerant turf varieties; and sensors, irrigation sprayers, and digital control systems that allow for more precise application of irrigation.

**Mining and industry.** Respondents representing mining and industrial water withdrawals commonly cited market demand for their product as a leading factor in water use at their facility. Customer demand for environmentally friendly products, technological improvements in process efficiency, reverse osmosis technology, and changing regulations were also noted.

**Thermoelectric power.** Some of the comments in the responses indicated that power utilities have significant confidence in their predictions of future water use. Power utilities withdraw, by far, the most water of any other water use category (DHEC, 2015). Interconnections in the power grid extend across the continent, buffering individual power plants from variations in local demand. Upgrades, renovations, or closures are generally the result of years of planning.

**Water supply.** Water suppliers commonly cited increased development of residential housing and commercial and industrial sector growth as drivers of water use. Some suppliers provide water to bottling plants, and growth in the market for bottled water is expected to continue. Factors that various suppliers mentioned that could decrease their consumptive withdrawals included state and federal regulations (including the progression of regulations allowing for aquifer recharge using treated wastewater), more efficient fixtures and building codes that mandate their use, water rate increases, increased industrial water reuse, reverse osmosis water treatment technology, leak-detection equipment, rainwater collection, automatic meter reading and advanced metering infrastructure (better ensuring accurate metering), and water for outdoor irrigation withdrawn directly from lakes by lakefront property owners .

## DISCUSSION

This project is an example of effective collaboration between South Carolina Department of Natural Resources, SCDHEC, and the permitted and registered water withdrawers in the state for the common goal of water resource management. The responses represent a good-faith effort by the various communities of water users in the state to provide valuable feedback for water planning. The project has achieved the specific goals introduced here, and it has done so at a very low cost to the state. Based on the results compiled in this effort, certain critiques can be made regarding the survey items and the interpretability of responses. Survey items about reporting accuracy, concerns about water availability, water reuse and reclamation, and qualitative responses are discussed further.

**Reporting accuracy.** The results indicate that reporting accuracy and precision varies between withdrawal categories and between methods of calculating withdrawal volume. In addition to the respondents' perceived accuracy, digitization of their handwritten withdrawal reports on the mandatory reporting forms can introduce additional error and uncertainty.

**Concerns about water availability.** The survey item regarding concerns about water availability (Figure 3) does not distinguish between physical and legal availability. Many withdrawers are not concerned about either, but for those who expressed concern, it is unclear whether their concern is based on the possibility of a drought or a groundwater decline causing a shortage or if the concern is more related

to perceived uncertainty in the water users' legal right to use their water supply.

**Water reuse and reclamation.** The survey item regarding water reuse and recycling was ambiguous and could be improved. On-site reuse is a water conservation practice that reduces both total withdrawal and total discharge for a given enterprise. In the context of water demand modeling, on-site reuse can be considered equivalent to other water conservation measures.

Reclamation of discharges, on the other hand, reduces the return flows from the contributing enterprise (the discharger) and helps to satisfy the water demands of the receiving enterprise. This practice can reduce total withdrawal, and it can also improve water quality if pollutants in the wastewater are diverted from environmental water bodies. In the context of water demand modeling, reclamation of discharges from one enterprise to another can be considered more akin to water distribution than water conservation. It was noted in response comments that reclaimed wastewater supplies can be impinged by water conservation at the contributing enterprise. In Texas, wastewater return flows are allocated similarly to other sources of water, and conservation efforts that could reduce return flows are subject to regulatory review to prevent or mitigate downstream shortages.

**Qualitative responses.** The responses described in the qualitative summary can be interpreted as a partial list of factors potentially affecting withdrawals in the different sectors. The relative importance of these different factors is subject to interpretation. These results can serve as a starting point for further investigation of technological and economic trends affecting water use in South Carolina. These responses are not expected to compose an exhaustive and definitive list, but rather a compilation of informed opinions.

## RECOMMENDATIONS FOR FUTURE WORK

The data collected in this work also allows for an analysis of water use specific to each withdrawal. The detailed responses will be used to calibrate models of water use based on the number of residential, commercial, or industrial taps (for water suppliers) and the acreage irrigated and crops planted (for agricultural and golf course withdrawals). The detailed responses from this survey will enable enterprise-level water use models that may be used for projecting future water demand.

The surveys could be improved by addressing the ambiguities discussed in the previous section. The survey forms could also be customized for each category of withdrawal. Although the survey could easily be repeated at a minimal cost next year, that could lead to a decline in the response rate, because water users' may not be motivated to respond to additional paperwork every year. Alternatively, a similar survey could be repeated in 2–5 years to monitor water users' changing perspectives over time. The response rate might be improved by offering an incentive such as a raffle for a free T-shirt or a subscription to *South Carolina Wildlife* magazine.

## LITERATURE CITED

Castro JE, Hu J. 1997. Distribution and rate of water use in South Carolina, 1994 (Hydrology–Water Resources Report 18). Columbia (SC): South Carolina Department of Natural Resources.

Harrigan JA. 1985. Water use in South Carolina, July–December, 1983 (Hydrology–Water Resources Commission Report 148). Columbia (SC): South Carolina Department of Natural Resources.

Holland TW. 1992. Water-use collection techniques in the Southeastern United States, Puerto Rico, and the U.S. Virgin Islands. (Water-Resources Investigations Report 92-4028). Little Rock (AR): U.S. Department of the Interior.

South Carolina Department of Health and Environmental Control (SCDHEC). 2012. R.61-119, Surface water withdrawal, permitting, and reporting. Columbia (SC): Author.

South Carolina Department of Health and Environmental Control (SCDHEC). 2017. 2015 Reported water use: South Carolina. Columbia (SC): Water Quantity Permitting Section, Division of Water Monitoring, Assessment, and Protection, Bureau of Water. https://www.scdhec.gov/sites/default/files/docs/ Environment/docs/WaterQualityStandards/2015%20 Reported%20Water%20Use.pdf.

Texas Water Development Board. 2017. Required survey of ground and surface water use for the calendar year ending December 31, 2017. Austin (TX): Author. https://www. twdb.texas.gov/waterplanning/waterusesurvey/survey/ doc/2017/2017_Coverletter.pdf.

## Appendix:
## Cover Letter and Water Use Survey

### South Carolina Department of
# Natural Resources

Alvin A. Taylor
**Director**

Ken Rentiers
Deputy Director for
**Land, Water and Conservation**

Greetings permitted or registered water withdrawer:

First and foremost, we would like to thank you for your compliance with South Carolina's water use permitting, registration, and reporting regulations. The information you provide is crucial to ensure adequate management of our State's water resources.

Our State is blessed with an abundance of water resources, to the benefit of many diverse stakeholders. However, intense and unpredictable weather extremes pose significant hazard, and our growing population and economy can increase our need for clean and reliable water supplies. Furthermore, we must plan and protect our water resource interests to mitigate potential conflicts within South Carolina and across state lines.

The South Carolina Department of Natural Resources has initiated a multi-year effort to develop Regional Water Plans for the State. Water demand forecasts can help all stakeholders and water users plan effectively. We recognize that water users have valuable insights from firsthand experience using water to meet their needs. We ask for your assistance in providing accurate data to improve our water demand forecast for your enterprise.

The attached survey includes 20 questions regarding water use at your enterprise. The questions are designed to get your input on the importance of your water sources, to understand the accuracy of reported water withdrawals, and to better understand how water is used at your enterprise. By participating in this survey, you can help ensure that the needs and interests of your organization are represented accurately in your basin's Regional Water Plan. Your response to this survey is optional and voluntary. Responses will be compiled and combined with other datasets to develop water demand models for public supply, irrigation, energy utilities and industry. These models will be reviewed by technical advisory committees and made available for review by the general public.

For FARMERS: Clemson will be conducting a more detailed and thorough survey in the Spring of 2018. If you provide your Registration ID in this survey, then your responses be shared with investigators at Clemson so that our data collection efforts are as efficient as possible.

For MUNICIPAL OR RURAL SUPPLIERS: We kindly request a digital map, shapefile, or geodatabase of your service area. If you sell water to other distributors, details regarding sales volume could significantly improve our water demand modelling results.

**Rembert C. Dennis Building • 1000 Assembly St • P.O. Box 167 • Columbia, S.C. 29202**

EQUAL OPPORTUNITY AGENCY                 www.dnr.sc.gov                 PRINTED ON RECYCLED PAPER ♻

## Instructions for responding to the survey

Questions 1-3 connect your survey responses to the other information we have regarding water use at your enterprise. The Permit/Registration ID code refers to the User ID on your Water Withdrawal Permit or Registration. The ID# is two numeric digits followed by two letters followed by three numeric digits (for example: 02WS045).

Questions 4-13 are multiple choice. We hope they are self-explanatory and easily answered. Questions 14 and 15 are to better understand your enterprise's water budget. Gathering data to estimate a detailed water budget can be costly and time consuming, but even a rough estimate could be informative.

Question 16 is for water suppliers. Although we do have some relevant information from withdrawal and distribution permits, some of that information is incomplete or out of date. Sales from one water supplier to another are particularly relevant for developing models of water demand.

Question 17 is designed for irrigators and water suppliers, but it can be used by any enterprise which keeps records of how monthly water withdrawal volume is used. The examples below illustrate how to use the table to describe water use at your enterprise:

Example 1. Agricultural Irrigation

| Year | Account / Crop | Taps / Acres | Jan | Feb | Mar | Apr | May | Jun |
|---|---|---|---|---|---|---|---|---|
| 2016 | Corn | 675 | - | - | 0.9 | 2.7 | 5.5 | 7.3 |
| 2016 | Soybeans | 800 | - | - | - | 2.2 | 4.3 | 6.5 |
| 2017 | Corn | 800 | - | - | 1.1 | 3.3 | 6.5 | 8.7 |
| 2017 | Soybeans | 675 | - | - | - | 1.8 | 3.7 | 5.5 |

Example 2. Municipal Water Supplier

| Year | Account / Crop | Taps / Acres | Jan | Feb | Mar | Apr | May | Jun |
|---|---|---|---|---|---|---|---|---|
| 2016 | Residential | 12,096 | 6.4 | 5 | 5.5 | 6.6 | 6.6 | 7.2 |
| 2016 | Commercial | 1,113 | 0.2 | 0.2 | 0.2 | 0.2 | 0.3 | 0.3 |
| 2016 | Wholesale | 1 | 2.1 | 2.0 | 2.1 | 2.7 | 2.9 | 3.0 |
| 2017 | Residential | 12,124 | 4.7 | 4.9 | 5.2 | 7.4 | 6.3 | 8.1 |

These examples use million gallons per month to fill in the table, as is required by DHEC for water withdrawal reporting. If you prefer to use another unit of measurement in your survey response, please note what unit you are using in the margin of the table. Information for a single year will be helpful in understanding current baseline water use. Information for multiple years will be helpful in developing statistical forecasts of future water use.

Questions 18 and 19 are open ended, and are intended to direct our research efforts to understand current developments and trends in water use in our State. Question 20 provides you with an opportunity to stay involved with water demand forecasting efforts. You can find more information and sign up for announcements at www.scwatermodels.com

Thank you in advance for your assistance with this study. If you have any questions, comments, or concerns regarding this survey or the water demand forecasts, please feel free to contact Alex Pellett using the contact information below. Please return survey responses to DHEC with your water withdrawal reporting form.

C. Alex Pellett
Hydrologist, SCDNR
Fax: (864) 654-9168
Phone: (864) 986-6255
Email: PellettC@dnr.sc.gov

1. Contact name_____

2. Enterprise name_____

3. Permit/Registration ID# _____

4. How important are your current water supplies for your enterprise?
   - o Critical – without current sources, enterprise would cease.
   - o Very important – could obtain water from other sources, at significant cost.
   - o Somewhat important – contingency plans minimize costs during a shortage
   - o Not important – enterprise does not rely on current water supplies

5. Are you concerned about future water availability for your enterprise?
   - o Not at all concerned
   - o Somewhat concerned
   - o Slightly concerned
   - o Very concerned

6. How precisely do you report monthly water withdrawals?
   - o Exactly correct
   - o Within 10%
   - o Within 20%
   - o Greater than 20% uncertainty

7. How do you calculate monthly water withdrawals?
   - o Flow meter
   - o Based on time pumping
   - o Based on energy spent pumping
   - o Best estimation/reckoning
   - o Other: _____

8. Over the next 5 years, I plan to:
   - o Increase water withdrawals
   - o Decrease water withdrawals
   - o Maintain the same volume of withdrawal
   - o Do not know

9. Over the next 5 years, I plan to:
   - o Use more surface water
   - o Use more groundwater
   - o Use more purchased water
   - o Maintain the current level of withdrawals
   - o Do not know

10. What factors will impact your water use in the future?
   - o Production practices reducing withdrawal
   - o Capital investments reducing withdrawal
   - o Production practices increasing withdrawal
   - o Capital investments increasing withdrawal

11. Are there any existing studies that forecast water demand for your enterprise?
   - o Yes
   - o No
   - o Do not know

12. Does your enterprise purchase water?
   - o Yes
   - o No

13. Does your enterprise practice water re-use or use reclaimed or treated wastewater?
   - o Yes
   - o No
   - o Do not know

14. What percent of water used at your enterprise is returned to groundwater?

15. What percent of water used at your enterprise is returned to a river, stream, lake, or pond?

16. Do you sell water to other water distributors? If so, to whom?

17. Please describe water use at your enterprise using the table below:

| Year | Account / Crop | Taps / Acres | Jan | Feb | Mar | Apr | May | Jun | Jul | Aug | Sep | Oct | Nov | Dec |
|---|---|---|---|---|---|---|---|---|---|---|---|---|---|---|
| 1 | | | | | | | | | | | | | | |
| 2 | | | | | | | | | | | | | | |
| 3 | | | | | | | | | | | | | | |
| 4 | | | | | | | | | | | | | | |
| 5 | | | | | | | | | | | | | | |
| 6 | | | | | | | | | | | | | | |
| 7 | | | | | | | | | | | | | | |
| 8 | | | | | | | | | | | | | | |
| 9 | | | | | | | | | | | | | | |
| 10 | | | | | | | | | | | | | | |
| 11 | | | | | | | | | | | | | | |
| 12 | | | | | | | | | | | | | | |
| 13 | | | | | | | | | | | | | | |
| 14 | | | | | | | | | | | | | | |
| 15 | | | | | | | | | | | | | | |
| 16 | | | | | | | | | | | | | | |
| 17 | | | | | | | | | | | | | | |
| 18 | | | | | | | | | | | | | | |
| 19 | | | | | | | | | | | | | | |
| 20 | | | | | | | | | | | | | | |
| 21 | | | | | | | | | | | | | | |
| 22 | | | | | | | | | | | | | | |
| 23 | | | | | | | | | | | | | | |
| 24 | | | | | | | | | | | | | | |
| 25 | | | | | | | | | | | | | | |

18. What technological developments do you believe will impact your water use in the future?

19. What market trends do you believe will impact your water use in the future?

20. Would you like to be included in email updates regarding this survey and other water forecasting efforts? If so, please include your email address below.

**PLEASE RETURN SURVEY RESPONSES TO DHEC WITH WATER WITHDRAWAL REPORTING FORMS**

Journal of South Carolina Water Resources, Volume 5, Issue 1, Pages 45–59, 2018

# South Carolina Groundwater Availability Assessment: 2017 Stakeholder Outreach and Engagement Results

Thomas Walker III[1], Lori Dickes[1], and Jeffery Allen[1]

AUTHORS: [1]South Carolina Water Resources Center, Clemson University, 509 Westinghouse Road, Pendleton, SC 29670.

**Abstract.** An update of the State Water Plan is underway in South Carolina. The purpose of the State Water Plan is to develop a water resources policy for South Carolina. A significant portion of the State Water Plan update is to include stakeholders into the planning process. Clemson University continues to facilitate the stakeholder engagement components of the steps to an updated water plan. This research is pertinent to the Groundwater Availability Assessment phase of the State Water Planning process. Overall, stakeholders were interested in all identified groundwater areas of interest in South Carolina. Additionally, they intended to be involved in the entire stakeholder process for groundwater and became more informed on the Groundwater Availability Assessment. Stakeholders agreed that groundwater modeling provided useful information for users in the state and thought the Groundwater Availability Assessment was important for water resources management. Nuances in stakeholder types and registered or permitted users versus nonregistered or nonpermitted users provide important details beyond general results. Moving forward, there are some more mixed results of the stakeholder engagement meetings that are important for planning and decision-making. The groundwater assessment meeting results had general agreement about the appropriateness of the scope, but had less certainty than other questions. Stakeholders generally identified the need for the allocation of additional resources for the planning process. Additionally, mixed results highlight the differences surrounding perceptions of the need for statewide permitting of groundwater resources. This exploratory research is important to water management in South Carolina because it assesses buy-in from those interested in or affected by water resource recommendations forthcoming at the end of the State Water Plan update.

## INTRODUCTION

The purpose of the South Carolina Groundwater Availability Assessment is to update the 2010 groundwater flow model of the coastal plain (Gellici, 2017). Groundwater flow models are useful in predicting water-level declines, recharge rates, and impacts of groundwater withdrawals on aquifers, streamflows, and other users in the coastal plain (South Carolina Department of Natural Resources [SCD-NR], 2015). Clemson University was contracted to facilitate the stakeholder engagement meetings as a part of the overall Groundwater Availability Assessment process. Because stakeholder involvement is a new approach to water planning in South Carolina, this study explored this approach taken in the Groundwater Availability Assessment pilot stakeholder engagement efforts.

The South Carolina Water Resources Center maintains an email distribution database with a third-party vendor for distribution of relevant water center news and events. The email distribution database is divided into several groups or data segments. Two population segments of the database were sent invitations to Phase 1 of the groundwater assessment meetings. The first group included 2,134 contacts, of which 494 invitations were opened (24.9%), whereas the second group had 369 identified groundwater specific stakeholders, of which 123 were opened (36.9%). This groundwater-specific stakeholder list was put together, in part, by the Stakeholder Engagement Team (SET), after calling 224 stakeholders to determine contact information for additional stakeholder communication. The SET is composed of South Carolina Water Resources Center researchers contracted to facilitate water planning stakeholder meetings throughout the state. The South Carolina Groundwater Association was also contacted to distribute the meeting information to their distribution list comprised of about 81 members. This was distributed on November 14, 2017, to all Groundwater Association members by its staff. Social media provided additional outreach: Twitter, Facebook, and LinkedIn.

Twitter posts had 2,269 impressions and 96 engagements. Facebook posts reached 675 viewers, with 14 post clicks and 9 reactions, comments, or shares; LinkedIn posts received approximately 90 views in the feed.

The Phase 1 Groundwater Availability Assessment stakeholder meetings were held on November 28, 2017, in North Charleston, South Carolina, and December 14, 2017, in West Columbia, South Carolina. The format of the meetings was as follows: (a) Clemson University welcome and introduction, (b) SCDNR water planning process update and groundwater methodology, (c) SCDNR Groundwater Availability Assessment, (d) U.S. Geological Survey (USGS) groundwater modeling, (e) Clemson University stakeholder engagement, (f) Clemson University concluding remarks, and (g) question-and-answer session with presenters facilitated by Clemson SET.

Although stakeholder engagement has become increasingly common (at times mandated) in the water planning process, there are different methodologies and approaches used in this type of process. Stakeholder involvement in the Groundwater Availability Assessment was desirable to move in a different direction from past top–down decision-making approaches in state water planning efforts. Stakeholder processes can range from informational to substantive stakeholder-driven decision-making (Cowie & Borrett, 2005). This informative-advisory process combines disseminating information with gathering information on stakeholder perceptions. This research highlights the results of a hybrid approach to stakeholder engagement, one that accounts for potentially larger groups but still gathers individual information and feedback. Additionally, this approach adds applied research into stakeholder perceptions of groundwater modeling and groundwater-specific resource and policy issues in South Carolina from various perspectives: general stakeholder perceptions, perceptions based on stakeholder type, and registered or permitted user vs. nonregistered or nonpermitted user perceptions.

## LITERATURE REVIEW

Collaborative resource management has become the new norm as water planning and policy has expanded beyond the purview of the state agencies charged with its management and regulation. As water planning and management have evolved in the United States, so has the collaborative nature of the water use sectors and stakeholders incorporated into the planning and decision-making process. Today, there is strong support for collaborative management as critical to successful decision-making processes and outcomes (Margerum & Robinson, 2015). Collaborative management can be time consuming, as parties involved generally have framed water management differently depending on their use, which can increase the time needed to build consensus (Margerum & Robinson, 2015).

Watershed partnerships with public involvement are an evolving area in water resource policy and management. Watershed partnerships go by many different names, such as councils, advisory groups, task forces, and committees (Sommarstrom in Born & Genskow, 1999). Even if the process or management of these groups varies, in general, they are local or regional groups of stakeholders who meet to discuss and collaborate on relevant water policy and management at a watershed (or portion of a watershed) scale (Leach & Pelky, 2001).

Groundwater modeling has become relatively standardized after decades of national and state modeling efforts. Literature has pointed to mixed results in participatory modeling including concerns about stakeholder involvement possibly degrading the scientific approach necessary to derive a quality product (Assata et al., 2008). Due to the well-defined and established methodology of assessing groundwater in South Carolina (Aucott et al., 1987; Campbell et al., 2017; Gellici, 2017) a fully participatory approach was not desirable for this phase of water planning. The question remains how the groundwater models will be used in future management and policy decision-making and enforcement.

Stakeholder engagement, therefore, is a complex issue. There have been a wide range of benefits documented as a result of stakeholder engagement efforts (Arnstein, 1969; Pretty, 2002; Ross et al., 2002; Rowe & Frewer, 2005; International Association for Public Participation, 2014). The benefits include, but are not limited to, more effective decision-making by public and private parties around complex issues, more transparency and knowledge sharing by public organizations, enhanced understanding by government agencies of the policy impacts on communities and individuals, and improved knowledge of governmental processes for individuals and organizations in the community (Newig, 2007; Mackenzie et al., 2012). Overall stakeholder processes provide an opportunity to improve information transfer, increase knowledge for all parties, build capacity, and create networking opportunities around complex issues.

## METHODOLOGY

Depending on project goals and objectives, as well as the time and resources available, stakeholder engagement can use different formats. For water management, stakeholders have a primary need for reliable, contemporary data on current groundwater availability, but also current data on withdrawals and use (Dilling et al., 2015). Given this, the process for these stakeholder engagement meetings followed more of an informational-advisory approach. Using Cowie and Borrett's (2005) model of decision-making, informational stakeholder engagement falls under a notification type of forum in which information is distributed and issues are explained; and advisory stakeholder engagement gathers perception feedback information.

In addition to information dissemination, real-time feedback was gathered by the Clemson SET using iClickers, and stakeholders were able to provide perception feedback and ask specific questions about the groundwater assessment project. iClickers are an information-collection tool used primarily in higher education to engage students in classroom settings. iClickers can also be used by researchers in other settings to provide anonymous feedback opportunities. This data collection technique was used to quantify stakeholder perceptions for more in-depth analysis in this process. The sampling performed for data collection and subsequent analysis followed a convenience sampling approach (Etikan et al., 2016).

The primary limitation of this engagement model is that, although it is efficient, it does not allow for building robust collaborative engagement. However, this model can be understood as a critical first step for complex issues that require building a foundation of information and networking from which to build a more collaborative and action oriented stakeholder processes.

## GROUNDWATER STAKEHOLDER ENGAGEMENT RESULTS

The Clemson SET chose to use iClicker polling as the primary information and engagement approach in the groundwater phase of the planning process to allow for anonymity and data aggregation. This was the same approach chosen for the initial phase of the surface water stakeholder engagement meetings throughout the 8 river basins across South Carolina from 2015 to 2016. Surface water hydrology and groundwater hydrogeology were differentiated in the planning process. SCDNR decided the surface water planning process would follow a basin approach, whereas groundwater would follow a coastal plain approach, divided into inner coastal and outer coastal geographical areas. Stakeholder engagement meetings were held in each of the coastal geographical areas. The results are presented from 3 primary perspectives: (a) by stakeholder response to questions, (b) by stakeholder type in response to questions, and (c) by registered or permitted stakeholder category in response to questions.

Appendixes 1–3 are the groundwater data sets used to create figures for the results section. The 2 meeting data sets were combined and then analyzed to provide a broader analysis of state groundwater stakeholder perceptions because limited analysis could be performed using the 2 sets individually. There are limitations to the results because they are not generalizable to all groundwater stakeholders due to the sampling methodology, which was a convenience

sampling approach rather than random sampling (Etikan et al., 2016). Additionally, a qualitative component might have provided more detailed responses. Sign-in sheets were examined for participant crossover to prevent double responses, and there was crossover in some governmental attendance but no crossover of meeting participants that provided feedback. The questions were asked to collect data on and document groundwater stakeholder demographics, groundwater interests, and stakeholder perceptions of the Groundwater Availability Assessment.

The first question for participants used a typology approach to categorize stakeholders (Appendix 1, Question 1). A typology of stakeholders can be important for policy and planning purposes. The typologies created for the groundwater assessment are broad due to the nature of the information collection tool. The types of stakeholders in water planning have many nuances that are difficult to capture with broad categories. However, categorization can be important to better understand perceptions and analyze feedback patterns of stakeholder responses based on type.

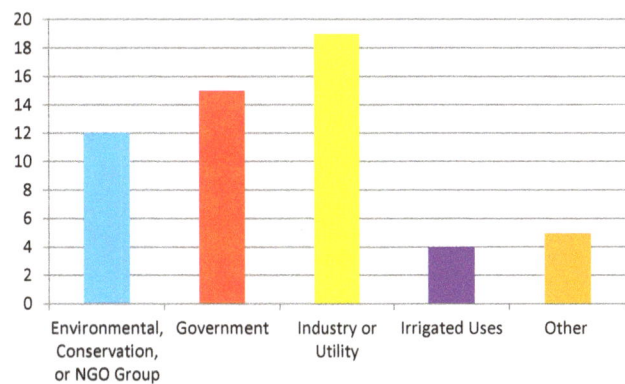

**Figure 1.** Organization type represented by stakeholder response. NGO = nongovernmental organization.

Industry and utility stakeholders had the strongest representation in the groundwater stakeholder meetings (Figure 1). Irrigated use stakeholders composed the least represented group in groundwater meetings. Government (local, state, federal, or higher education) and environmental, conservation, or nongovernmental organization (NGO) groups were also highly representative of the stakeholders in attendance at the groundwater meetings. Of interest is how irrigated use stakeholders are the least represented group but are identified as one of the main reasons water policy and legislation have been a point of emphasis in the state, largely due to perceptions of high and unregulated irrigation use (*Jowers et al. v. South Carolina Department of Health and Environmental Control [SCDHEC]*, 2018). Utility, industry, and irrigation stakeholders are more likely to be registered or permitted users versus environmental and government stakeholders (Figure 2; Appendix 3, Question 1).

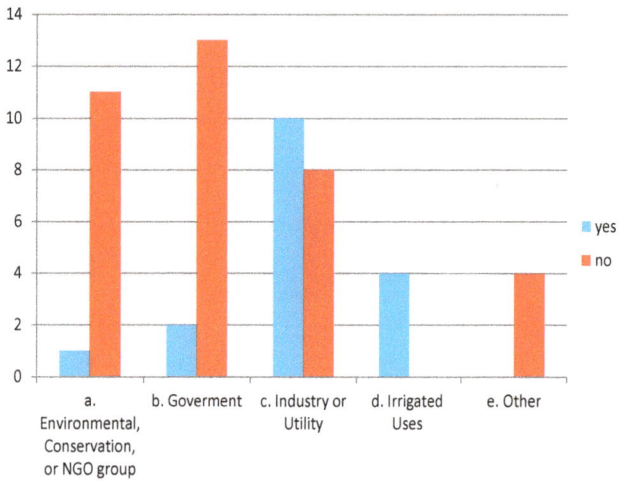

**Figure 2.** Organization type by registered or permitted stakeholder response. NGO = nongovernmental organization.

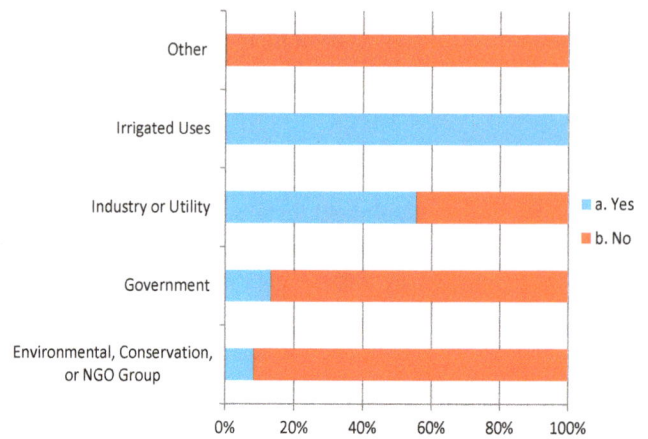

**Figure 4.** Registered or permitted response by stakeholder type. NGO = nongovernmental organization.

The next question directly asked if stakeholders were registered or permitted groundwater users (Appendix 1, Question 2). SCDHEC uses registration and permitting to account for groundwater use in the state of South Carolina. Registration is required if the user pumps 3 million gallons per month or more outside of the coastal plain; a notice of intent is required for groundwater pumping at or above the threshold in the coastal plain but not in a capacity use area. Permitting is required in capacity use areas of the state at or above the threshold (SCDHEC, 2018; Figure 3), and the state is currently in the process of designating a western South Carolina capacity use area in 2018. The majority

of stakeholders (67.9%) were not registered or permitted groundwater users in the state of South Carolina. This unequal representation could lead to nonrepresentative results. Results based on this response were analyzed to examine this possibility and are presented in subsequent figures. Industry or utility and irrigated uses composed the majority of the registered or permitted stakeholders, whereas there was a smaller percentage of government and environmental, conservation, or NGO registered or permitted stakeholders represented (Figure 4; Appendix 2, Question 1). Many other, government, environmental, conservation, and NGO groups were not registered or permitted groundwater users, with the industry or utility sector having almost equal representation

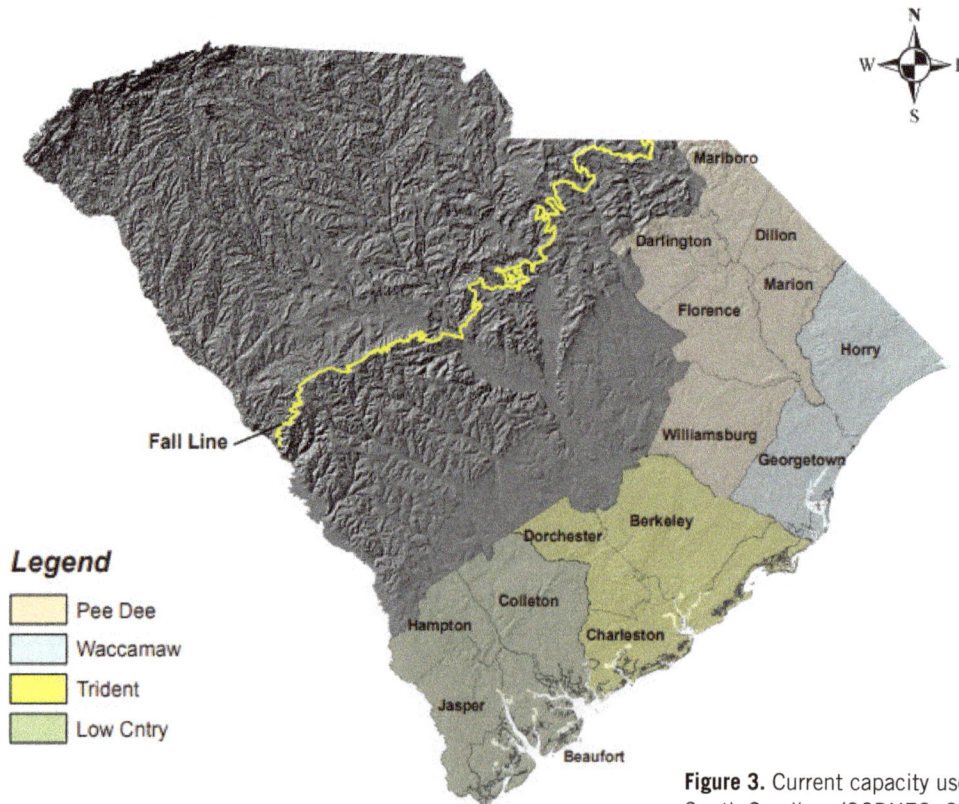

**Figure 3.** Current capacity use areas of South Carolina. (SCDHEC, 2018)

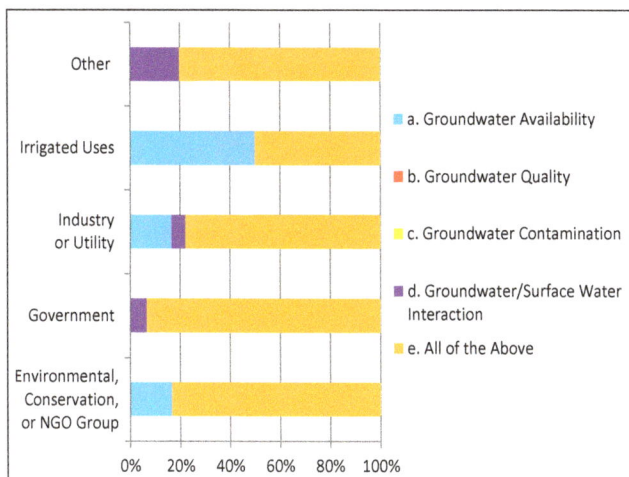

**Figure 5.** Stakeholder interests in groundwater by type. NGO = nongovernmental organization.

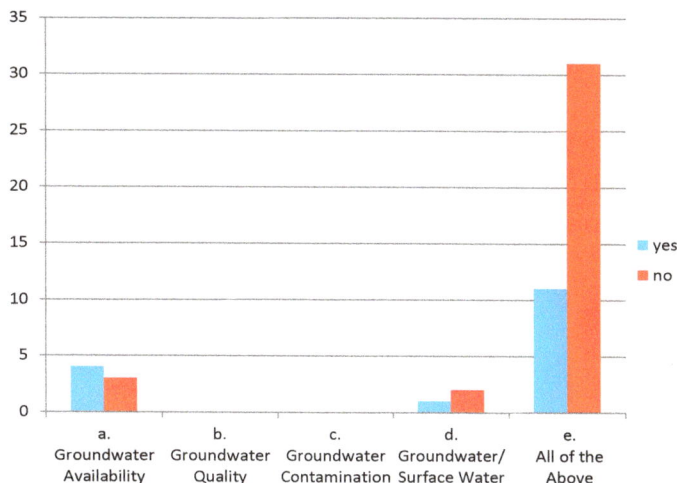

**Figure 6.** Groundwater interests by registered or permitted stakeholder response.

of registered or permitted users versus nonregistered or nonpermitted users.

The next question asked stakeholders to gauge their particular interest in groundwater (Appendix 1, Question 3). Although the purpose of the meetings was to inform stakeholders of the modeling efforts to assess the availability for current and future use in the state, very few were interested only in groundwater availability. Stakeholders were highly interested in all areas of groundwater issues in the state: availability, quality, contamination, and groundwater or surface water interaction (81.4%). Groundwater quality focuses on dissolved chemicals and gases in the water from the area geography (water quality can be poor and not be contaminated). Groundwater contamination focuses on more manmade applied or leaked chemicals seeping into the groundwater. All of these (groundwater or surface water availability, groundwater quality, groundwater contamination, and groundwater or surface water interaction) had the highest response rate regardless of stakeholder type (Figure 5; Appendix 2, Question 2). The irrigated use stakeholder group was almost evenly split between interests in groundwater availability only or all of the topics, and some industry and utility and environmental, conservation, or NGO stakeholders also had interest in availability, in addition to all of these topics.

Some government, industry or utility, or other stakeholders had interest in groundwater or surface water interactions, as well as all other topics. The data revealed that registered or permitted stakeholders are generally most interested in all groundwater topics, but also showed an interest in just groundwater availability and groundwater or surface water interactions (Figure 6; Appendix 3, Question 2).

An important component of stakeholder engagement is classifying the geographical representation of participants. There were equal numbers of representatives from the Inner and Outer Coastal Plain areas of the state (Figure 7). The

most highly represented response was that stakeholders embodied all groundwater areas of the state. The least represented area of the state is outside of the coastal plain, which includes the Piedmont area and is not a part of the Groundwater Availability Assessment effort. Industry or utility and government stakeholders had the highest response for representing all groundwater areas of the state (Figure 8; Appendix 2, Question 3). Irrigated uses and environmental, conservation, or NGO stakeholders largely represent the Inner Coastal Plain. The Inner Coastal Plain has been an area of interest at the regulatory level of the state. Currently, SCDHEC is assessing expanding designated capacity use areas to include parts of western South Carolina due to groundwater pumping, affecting recharge rates. The Outer Coastal Plain had more representation from government, industry or utility, and other stakeholder groups. Even though the Piedmont area of the state is not currently part of the groundwater availability assessment, some stakeholder represented this area at the groundwater stakeholder

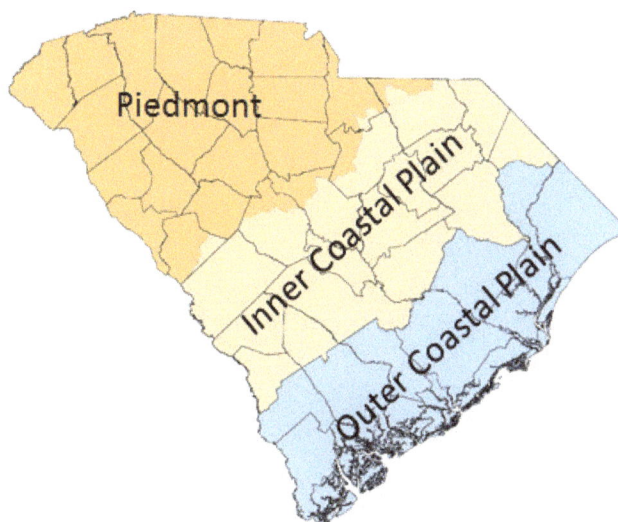

**Figure 7.** Geographical groundwater areas of South Carolina.

meetings from industry or utility and other stakeholder groups. Nonregistered or nonpermitted users are more likely to respond that they represent the entire state than registered or permitted users, according to data collected. Industry or utility and government stakeholders had the highest response rates for representing all groundwater areas of the state (Figure 8; Appendix 2, Question 3), and government had the highest response of nonregistered or nonpermitted stakeholders (Figure 2; Appendix 3, Question 1).

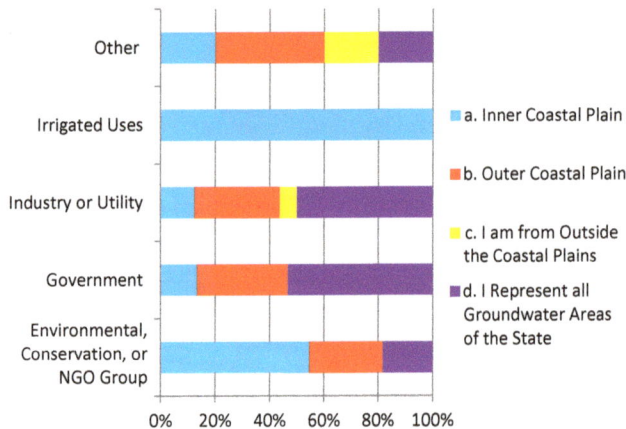

**Figure 8.** Geographical groundwater area represented by stakeholder type. NGO = nongovernmental organization.

A key objective of water planning and engagement is to support and encourage ongoing stakeholder involvement in these processes. The next question asked whether or not stakeholders intend to be involved in the entire groundwater stakeholder process (Appendix 1, Question 5).

Almost 75% of stakeholders who participated in the first round of groundwater engagement meetings responded that they will continue to participate in the entire process for the Groundwater Availability Assessment. Most stakeholders responded with interest in remaining engaged throughout the entire groundwater process in South Carolina (Figure 9; Appendix 2, Question 4). Additionally, there were some

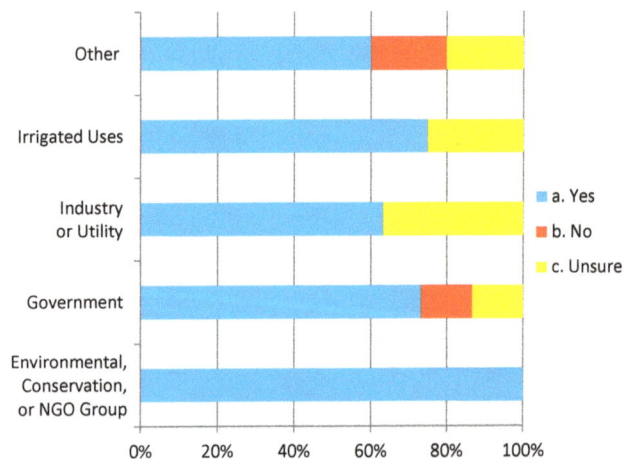

**Figure 9.** Intent to remain involved in the entire process for groundwater by stakeholder type. NGO = nongovernmental organization.

groups who were unsure if they would remain involved in the participatory process. Other and government stakeholder types indicated that they would not continue to be engaged after the first round of meetings. Registered or permitted stakeholders are more likely to remain engaged in the entire process (Figure 10; Appendix 3, Question 4).

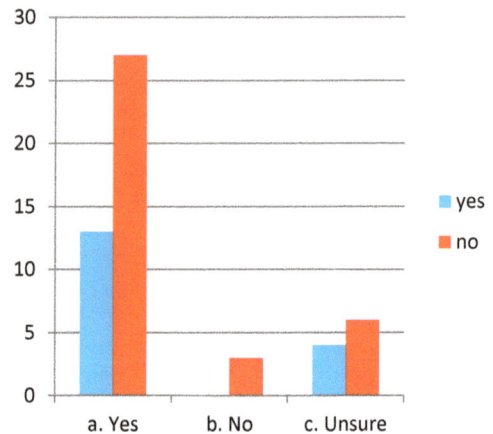

**Figure 10.** Intent to remain involved in the entire process for groundwater by registered or permitted stakeholder response.

The next several questions focused on the scope and resources of the groundwater availability and whether it was perceived as appropriate (Appendix 1, Questions 6 and 7). Stakeholders were asked these questions after presentations from SCDNR and USGS on the methodology, approach, and scope of the assessment. Approximately 80% of stakeholders strongly agreed or agreed that the scope of the Groundwater Availability Assessment was appropriate (Figure 11; Appendix 1, Question 6). Several groups did not know if the scope was appropriate, which could be due to a lack of familiarity with groundwater modeling at the state level or in general. There was some disagreement in industry or utility and environmental, conservation, or NGO groups and strong

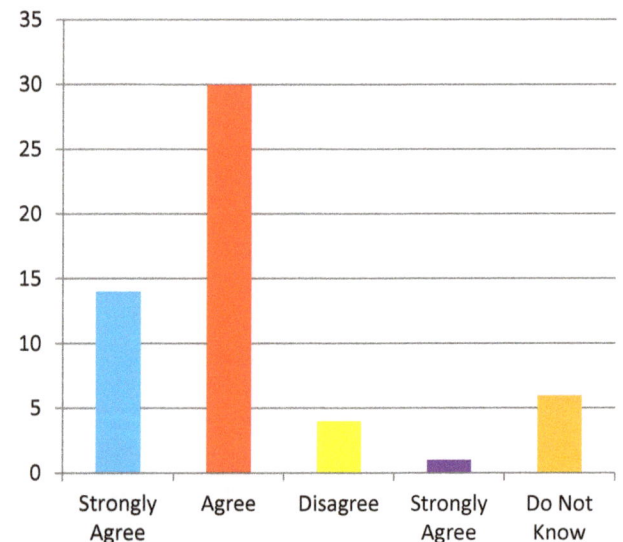

**Figure 11.** The scope of the Groundwater Availability Assessment is appropriate.

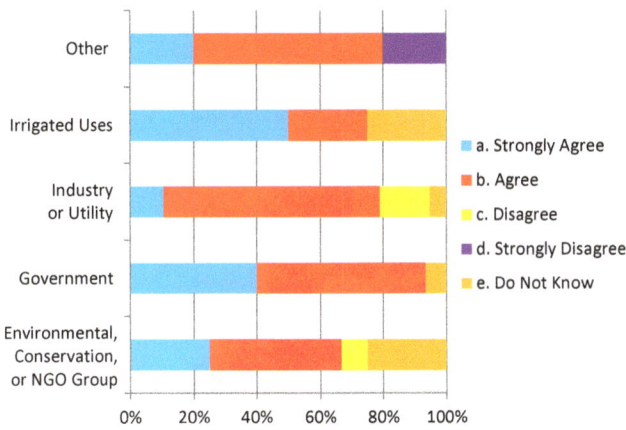

**Figure 12.** The scope of the Groundwater Availability Assessment is appropriate by stakeholder type. NGO = nongovernmental organization.

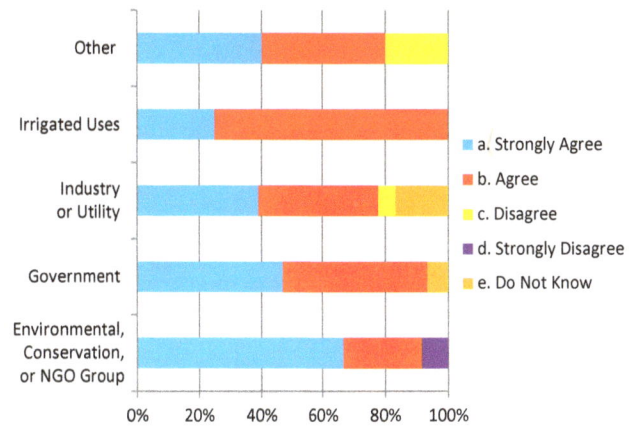

**Figure 13.** Support additional resources for the Groundwater Availability Assessment and planning efforts by stakeholder type. NGO = nongovernmental organization.

disagreement in the other stakeholder group (Figure 12; Appendix 2, Question 5). One of the more important aspects of stakeholder engagement for resource management and planning efforts is gauging approval of the scope of the project. Stakeholders generally agreed more than strongly agreed that the scope of the Groundwater Availability Assessment was appropriate regardless of registered or permitted stakeholder status. Both registered or permitted and nonregistered or nonpermitted stakeholders had some level of disagreement and some did not know if the scope was appropriate. These results are important for SC government and regulatory agencies as the state water planning process moves forward.

The next question focused more specifically on resources by addressing if stakeholders supported additional resource allocation for the state Groundwater Availability Assessment and water planning efforts (Appendix 1, Question 7). Support for additional resources for groundwater assessment and planning efforts received strong agreement. "Strongly agree" and "agree" received over 85% of the responses from stakeholders. This response is significant, but comes without discussion or information pertaining to the amount of resources currently being allocated to these efforts other than what was allocated for the surface water availability assessment efforts from the state legislature. The amount spent on the groundwater assessment was not disclosed in this round of stakeholder engagement. If this information was disclosed and stakeholders had a reference to gauge their response, their responses may be more nuanced than these results indicate. This would be especially important for those that responded "do not know." Additional reference points like other states' planning efforts and alternative modeling approaches were not discussed, nor was the amount of resources other states, especially neighboring states, have allocated to these efforts. Environmental, conservation, or NGO stakeholders and other had some disagreement in

their support for additional resources for the Groundwater Availability Assessment (Figure 13; Appendix 2, Question 6).

There was general consensus among stakeholders regardless of registered or permitted user status in support of additional resources for groundwater assessment and planning efforts. There was a slight difference between nonregistered or nonpermitted stakeholders and registered or permitted stakeholders in strength of agreement (Figure 14; Appendix 3, Question 6). Nonregistered or nonpermitted users strongly agreed with the statement more so than registered or permitted users, who agreed more than strongly agreed with the statement.

The next question gauged the strength of the information content of the presentations (Appendix 1, Question 8). Stakeholders strongly agreed or agreed that the information communicated in the stakeholder engagement meetings was informative. These results reveal that the information

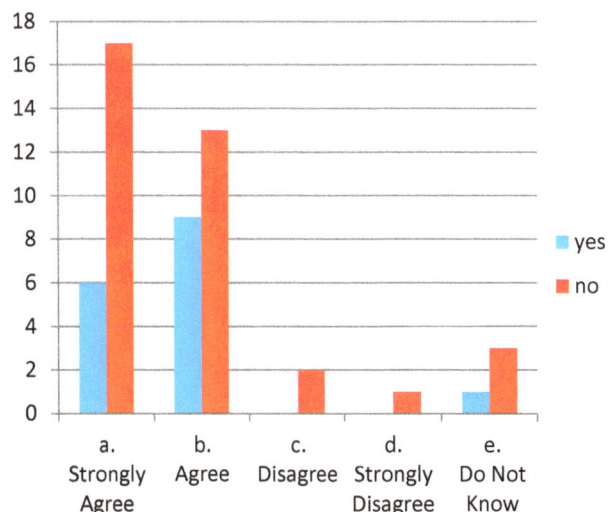

**Figure 14.** Support additional resources for the Groundwater Availability Assessment and planning efforts by registered or permitted stakeholder response.

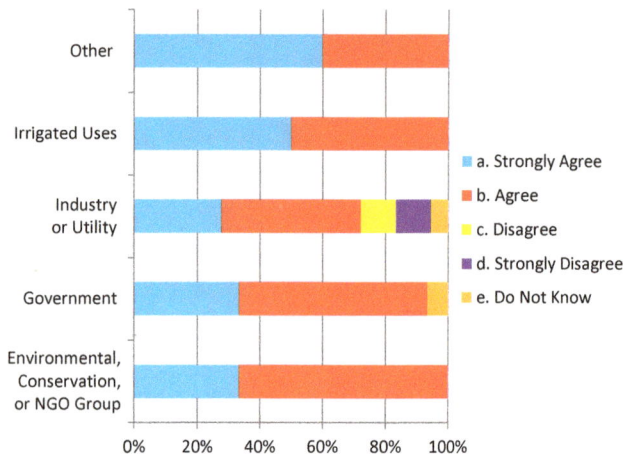

**Figure 15.** Stakeholders are more informed on the Groundwater Availability Assessment by stakeholder type. NGO = nongovernmental organization.

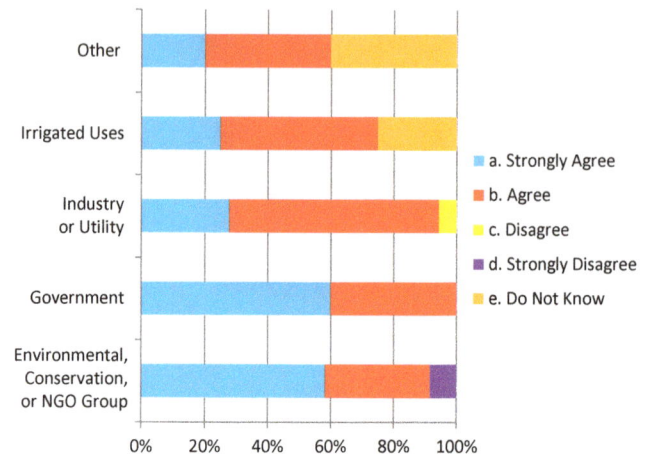

**Figure 16.** Groundwater modeling provides useful information for groundwater users in the state by stakeholder type. NGO = nongovernmental organization.

was presented in a manner that was understood by a diverse stakeholder group with various levels of expertise. The other stakeholder type strongly agreed more than the rest of the types that they were more informed on the Groundwater Availability Assessment from participating in the engagement meeting. All groups predominantly agreed to some extent that they felt more informed after the presentations from Clemson SET, SCDNR, and USGS. Industry or utility stakeholder groups had some disagreement and strong disagreement with the statement or did not know (Figure 15; Appendix 2, Question 7). Registered or permitted (81.25%) or nonregistered or nonpermitted (91.66%) stakeholders had high combined levels of "strongly agree" or "agree." Although we had some responses in the "disagree," "strongly disagree," or "do not know categories," those responses were very low.

The next question continued to inquire whether stakeholders were more informed after the Groundwater Availability Assessment presentations by asking if groundwater modeling provides useful information for groundwater users (Appendix 1, Question 9). In response, approximately 90% of stakeholders strongly agreed or agreed. The majority of stakeholder responses indicated that groundwater modeling efforts did provide useful information for groundwater users in the state. Industry or utility and environmental, conservation, or NGO both responded with some disagreement (Figure 16; Appendix 2, Question 8). Irrigated uses and other had some "do not know" responses, which could be reflective of the technical nature of groundwater modeling. There was little detectable difference between registered or permitted and nonregistered or nonpermitted stakeholders. However, there were a few responses from nonregistered or nonpermitted stakeholders that disagreed or strongly disagreed with the usefulness of this information.

As with many environmental planning efforts, one of the areas of ongoing concern for many stakeholders is how this information will be used and whether there is a

regulatory purpose for it. As such, stakeholders were asked whether or not they supported statewide groundwater withdrawal permitting. With the preliminary assessment and engagement for a new capacity use area in western South Carolina underway, groundwater registration and permitting policy has been receiving increased attention in the state. This question gauged stakeholders' general support for statewide groundwater permitting. Overall, the highest response from stakeholders was that they strongly support statewide groundwater permitting (54.7%). When looking at the overall distribution of responses, approximately 45% were not strongly in favor of statewide groundwater withdrawal permitting but, instead, responded that they did not support statewide groundwater permitting (3.7%), supported only regional permitting where groundwater problems exist (24.5%), or were not sure (16.9%; Figure 17; Appendix 1, Question 10).

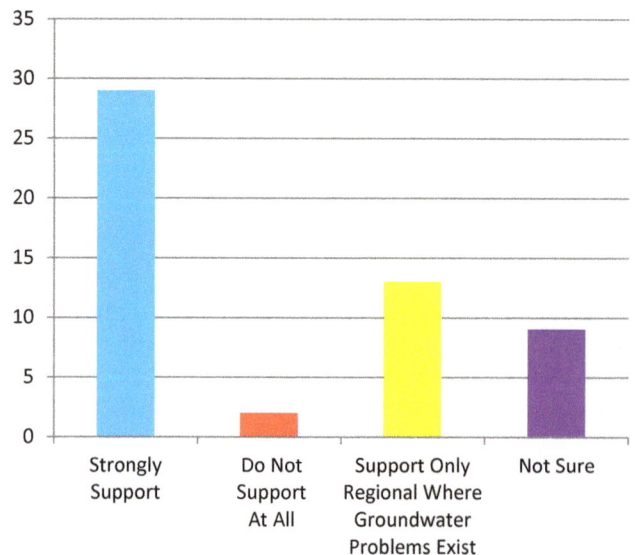

**Figure 17.** Stakeholder support for statewide groundwater withdrawal permitting.

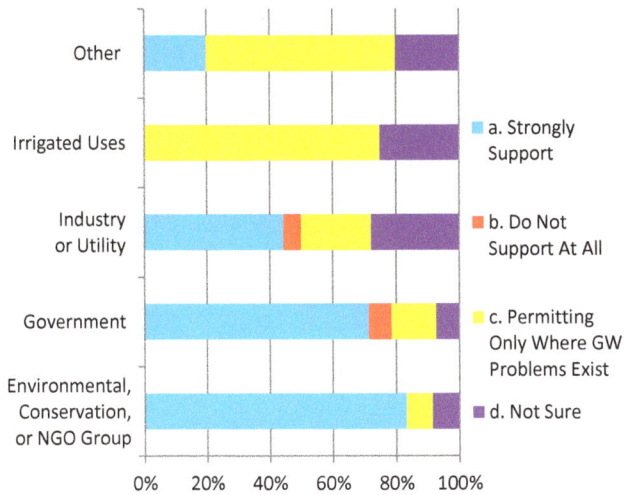

**Figure 18.** Stakeholder support for statewide groundwater withdrawal permitting by stakeholder type. NGO = nongovernmental organization.

Groundwater permitting across the state was most highly supported by environmental, conservation, or NGO groups, followed by government stakeholders (Figure 18; Appendix 2, Question 9). Industry and utility were more evenly distributed in their support. Irrigated uses and other stakeholders were more supportive of permitting only where identified groundwater problems exist or were uncertain. The stakeholder responses for support of various permitting options for the state were different for registered or permitted users and nonregistered or nonpermitted users. Registered or permitted users were well distributed among strongly supporting statewide groundwater permitting, supporting groundwater permitting only in regions where problems exist, and not sure if they support any of the available options (Figure 19; Appendix 3, Question 9). Nonregistered or nonpermitted stakeholders had higher response rates in

strongly supporting statewide groundwater permitting, but also had responses in not supporting statewide groundwater permitting, supporting groundwater permitting only in regions where problems exist, and not being sure if they support any of the available options. A high degree of variability in these responses is cause for reflection for policymakers. If any permitting efforts were proposed, a higher level of engagement and information sharing would be important. Currently, the state is undertaking a preliminary assessment for expanding designated capacity use areas registration and permitting, which, based on this question, provides an opportunity for the state to build capacity around this issue with appropriate investment in public engagement and information sharing.

The final question asked stakeholders how important they felt the Groundwater Availability Assessment is for water resources management in the state of South Carolina (Appendix 1, Question 11). The Groundwater Availability Assessment is a critical component for South Carolina to plan and ensure long-term access to the state's groundwater resources. Stakeholders highly agreed that this effort was either "very important" or "somewhat important" (81.4% and 14.8%, respectively). Therefore, the Groundwater Availability Assessment is important to all stakeholder groups. Irrigated use and environmental, conservation, and NGO stakeholders responded that the assessment was "very important" at a higher percentage than other stakeholder groups (Figure 20; Appendix 2, Question 10). Both registered or permitted and nonregistered or nonpermitted stakeholders responded similarly in that groundwater modeling efforts for the state were "very important" or "important." Very few stakeholders came away from the engagement meetings not knowing if groundwater modeling was important for state water resource planning.

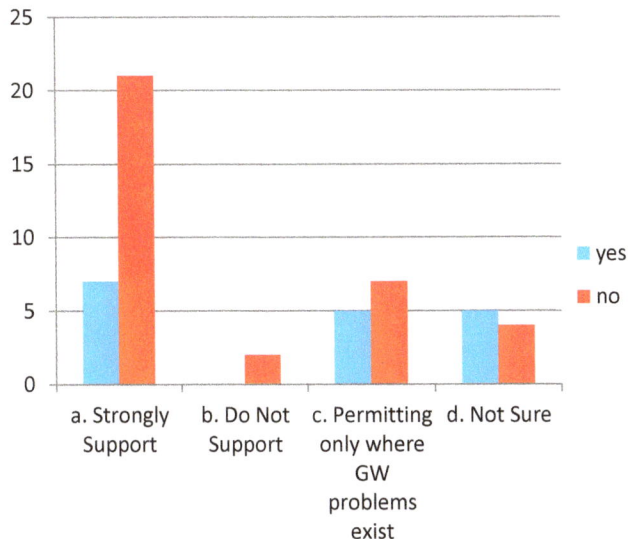

**Figure 19.** Stakeholder support for statewide groundwater withdrawal permitting by registered or permitted stakeholder response.

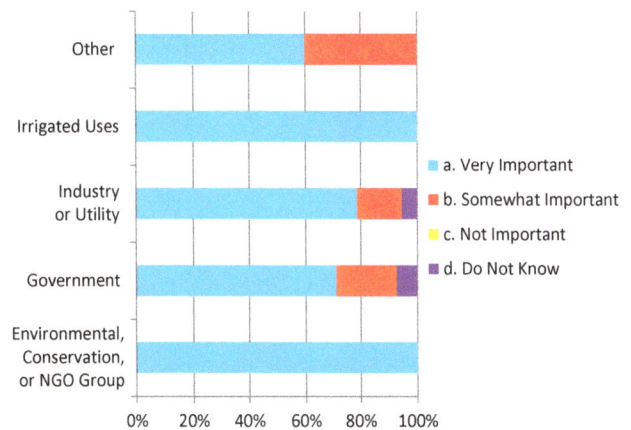

**Figure 20.** Importance of the Groundwater Availability Assessment by stakeholder type. NGO = nongovernmental organization.

## DISCUSSION

This is the first time the state of South Carolina has engaged in statewide water modeling and stakeholder engagement efforts around state water resource issues. As such, it is important to begin to characterize the nature of these types of efforts across the state. For example, who participates in these efforts and what is the nature of participation from different groups and organizations across the state? As well, it is increasingly important for policymakers and regulators to understand stakeholder perceptions around environmental and natural resource issues for effective management and use of these resources. Stakeholder methodology around issues that affect an entire region or state underscore the importance of broad and varied participation across groups. As such, these initial stakeholder meetings will allow the SET to begin to understand the potential gaps or weaknesses in participation. One such gap in participation could be agricultural interests. Agricultural water use and community perceptions of this use have been a point of contention in some areas of the state in that it is perceived that the agricultural industry is being treated differently than other water use sectors (*Jowers et al. v. SCDHEC*, 2018). Getting agricultural stakeholders engaged in the water planning process could perhaps reframe those perceptions if they were present to engage in meaningful dialogue with other stakeholders. Ensuring equity in participation is an important goal for this type of stakeholder effort. To ensure robust stakeholder efforts, they should incorporate methods of engagement, meeting style, and promotion that facilitate and support the broadest participation. This research reveals strong participation from key groups, with some responses highlighting areas of future research and stronger engagement needed in the future.

For example, several questions highlight areas of uncertainty where ongoing stakeholder involvement is important. The method of stakeholder involvement used here is valid and useful but is largely focused on providing information and understanding, as opposed to action-oriented processes. Results reveal there are several areas in which stakeholders have some degree of uncertainty in this process. Three areas specifically provide areas of consideration for policymakers in the state: (a) appropriateness of modeling efforts, (b) allocation of necessary resources, and (c) nature of potential permitting around groundwater resources.

All of these areas had enough stakeholders who indicated disagreement or uncertainty that opportunities exist for further information sharing or more in-depth engagement around these issues. For highly technical issues, like water modeling, ensuring that stakeholders have enough information, without creating more confusion or misunderstanding, is critical.

Similarly, any policy issue that may impact permitting and regulation necessitates a more robust stakeholder process.

Without additional engagement efforts, policymakers may find themselves in more contentious and challenging situations with different stakeholder groups as they attempt to implement policy changes. For example, models of stakeholder engagement that provide opportunities for different scenarios and dialogue around when additional permitting may be important could be useful in the future. As these results highlight, not all regulation is perceived negatively. However, additional information and dialogue from diverse groups would be critical to ensuring new regulation was perceived more positively.

Models of stakeholder engagement vary with a range of types of engagement and related outcomes (Cowie & Borrett, 2005). For the type of engagement initiated in this research, the outcomes were achieved. Results overwhelmingly reveal that individuals felt informed, had a greater understanding of the process, and were potentially primed for additional engagement. As the state moves forward developing advisory basins, more in-depth stakeholder efforts will be important. As noted, this is especially true around issues of permitting and regulations. In particular, stakeholder efforts that create conversation, commitment, and collaborative planning are important. Taking lessons from community-based resource management to increase knowledge and support for collaborative management of the state's water resources is a valuable approach for all stakeholders and one that could provide long-term benefits for the state and its natural resources.

## LITERATURE CITED

Arnstein SR. 1969. A ladder of citizen participation. J Am Inst Plann. 35(2):216–224.

Assata H, van Beek E, Borden C, Gijsbers P, Jolma A, Kaden S, Kaltofen M, Labadie JW, Loucks DP, Quinn NWT, et al. 2008. Generic simulation models for facilitating stakeholder involvement in water resources planning and management: a comparison, evaluation, and identification of future needs. Dev. Integr. Environ. Assess. 3:229–246.

Aucott WR, Davis ME, Speiran GK. 1987. Geohydrologic framework of the coastal plain aquifers of South Carolina (Water-Resources Investigations Report 85-4271). Columbia (SC): U.S. Geological Survey.

Born SM, Genskow KD. 1999. Exploring the watershed approach: critical dimensions of state–local partnerships: the Four Corners Watershed Innovators Initiative final report. Portland (OR): River Network.

Campbell B, Cherry G, Fine J, Butler A, Gellici J. 2017. South Carolina coastal plain groundwater availability model–an update. Presented at: Groundwater Availability Assessment–Meeting 1, November 28, 2017, North Charleston, SC.

Cowie GM, Borrett SR. 2005. Institutional perspectives on participation and information in water management. Environ Model Softw. 20:469–483.

Dilling L, Lackstrom K, Haywood B, Dow K, Lemos MC, Berggren J, Kalafatis S. 2015. What stakeholder needs tell us about enabling adaptive capacity: the intersection of context and information provision across regions in the United States. Weather Clim Soc. 7(1):5–17.

Etikan I, Musa SA, Alkassim RS. 2016. Comparison of convenience sampling and purposive sampling. Am J Theor Appl Stat. 5(1):1–4. doi:10.11648/j.ajtas.20160501.11.

Gellici J. 2017. The hydrogeologic framework developed for the South Carolina coastal plain groundwater flow model. Presented at: Groundwater Availability Assessment–Meeting 1, November 28, 2017, North Charleston, SC.

International Association for Public Participation. (2014). IAP2 Spectrum of Public Participation. https://cdn.ymaws.com/www.iap2.org/resource/resmgr/foundations_course/IAP2_P2_Spectrum_FINAL.pdf

Jowers JJ Sr, Anastos AJ, Williamson B, Ruhlman M, Ruhlman A v. SCDHEC. South Carolina Supreme Court. Appellate Case No. 2016-000428. 2018.

Leach WD, Pelkey NW. 2001. Making watershed partnerships work: a review of the empirical literature. J Water Resour Plan Manag. 127(6)378–385.

Mackenzie J, Poh-Ling T, Hoverman S, Baldwin C. 2012. The value and limitations of participatory action research methodology. J Hydrol. 474, 11-21.

Margerum RD, Robinson CJ. 2015. Collaborative partnerships and the challenges for sustainable water management. Curr Opin Environ Sustain. 12:53–58.

Newig J. 2007. Does public participation in environmental decisions lead to improved environmental quality? Towards an analytical framework. Int J Sustain Commun. 1(1):51–71.

Pretty J. 2002. Agri-culture: reconnecting people, land, and nature. London (UK): Earthscan.

Ross H, Buchy M, Proctor W. 2002. Laying down the ladder: a typology of public participation in Australian natural resource management. Aust J Environ Manag. 9(4):205–217.

Rowe G, Frewer LJ. 2005. A typology of public engagement mechanisms. Sci Technol Hum Values. 30(2):251–290.

South Carolina Department of Health and Environmental Control (SCDHEC). 2018. Groundwater use and reporting in capacity use areas–overview. Web.

South Carolina Department of Natural Resources (SCDNR). 2015. South Carolina surface-water quantity models: step 2 groundwater flow models. Presentation.

## APPENDIX 1:
## GROUNDWATER STAKEHOLDER ENGAGEMENT RESPONSES

| Question | n | % |
|---|---|---|
| 1. Which organization type do you represent? *n* = 55 | | |
| Environmental, conservation, or nongovernmental organization | 12 | 0.21818 |
| Government | 15 | 0.27273 |
| Industry or utility | 19 | 0.34545 |
| Irrigated uses | 4 | 0.07273 |
| Other | 5 | 0.09091 |
| 2. Are you a registered/permitted water user? *n* = 53 | | |
| Yes | 17 | 0.32075 |
| No | 36 | 0.67925 |
| 3. My interests are mainly in ___. *n* = 54 | | |
| Groundwater availability | 7 | 0.12963 |
| Groundwater quality | 0 | 0 |
| Groundwater contamination | 0 | 0 |
| Groundwater/surface water interaction | 3 | 0.05556 |
| All of the above | 44 | 0.81481 |
| 4. Which groundwater area do you represent? *n* = 51 | | |
| Inner coastal plain | 15 | 0.29412 |
| Outer coastal plain | 15 | 0.29412 |
| I am from outside the coastal plains | 2 | 0.03922 |
| I represent all groundwater areas of the state | 19 | 0.37255 |
| 5. Do you intend to be involved in the entire stakeholder process for groundwater? *n* = 55 | | |
| Yes | 41 | 0.74545 |
| No | 3 | 0.05455 |
| Unsure | 11 | 0.2 |
| 6. In my opinion, the scope of the Groundwater Availability Assessment is appropriate. *n* = 55 | | |
| Strongly agree | 14 | 0.25455 |
| Agree | 30 | 0.54545 |
| Disagree | 4 | 0.07273 |
| Strongly agree | 1 | 0.01818 |
| Do not know | 6 | 0.10909 |

| Question | n | % |
|---|---|---|
| 7. I support additional resources for the Groundwater Availability Assessment and planning efforts. *n* = 54 | | |
| Strongly agree | 25 | 0.46296 |
| Agree | 22 | 0.40741 |
| Disagree | 2 | 0.03704 |
| Strongly agree | 1 | 0.01852 |
| Do not know | 4 | 0.07407 |
| 8. Would you agree that you are now more informed on the Groundwater Availability Assessment? *n* = 54 | | |
| Strongly agree | 19 | 0.35185 |
| Agree | 29 | 0.53704 |
| Disagree | 2 | 0.03704 |
| Strongly agree | 2 | 0.03704 |
| Do not know | 2 | 0.03704 |
| 9. Groundwater modeling provides useful information for groundwater users in the state. *n* = 54 | | |
| Strongly agree | 23 | 0.42593 |
| Agree | 26 | 0.48148 |
| Disagree | 1 | 0.01852 |
| Strongly agree | 1 | 0.01852 |
| Do not know | 3 | 0.05556 |
| 10. Do you support statewide groundwater withdrawal permitting? *n* = 53 | | |
| Strongly support | 29 | 0.54717 |
| Do not support at all | 2 | 0.03774 |
| Support only regional where groundwater problems exist | 13 | 0.24528 |
| Not sure | 9 | 0.16981 |
| 11. How important do you feel the Groundwater Availability Assessment is for water resources management? *n* = 54 | | |
| Very important | 44 | 0.81481 |
| Somewhat important | 8 | 0.14815 |
| Not important | 0 | 0 |
| Do not know | 2 | 0.03704 |

APPENDIX 2:
GROUNDWATER STAKEHOLDER ENGAGEMENT RESPONSES BY ORGANIZATION TYPE

| Question | Environmental, Conservation, or Nongovernmental Group | Government | Industry or Utility | Irrigated Uses | Other |
|---|---|---|---|---|---|
| 1. Are you a registered/permitted water user? *n* = 53 | | | | | |
| Yes | 1 | 2 | 10 | 4 | 0 |
| No | 11 | 13 | 8 | 0 | 4 |
| 2. My interests are mainly in _____. *n* = 54 | | | | | |
| Groundwater availability | 2 | 0 | 3 | 2 | 0 |
| Groundwater quality | 0 | 0 | 0 | 0 | 0 |
| Groundwater contamination | 0 | 0 | 0 | 0 | 0 |
| Groundwater/surface water interaction | 0 | 1 | 1 | 0 | 1 |
| All of the above | 10 | 14 | 14 | 2 | 4 |
| 3. Which groundwater area do you represent? *n* = 51 | | | | | |
| Inner coastal plain | 6 | 2 | 2 | 4 | 1 |
| Outer coastal plain | 3 | 5 | 5 | 0 | 2 |
| I am from outside the coastal plains | 0 | 0 | 1 | 0 | 1 |
| I represent all groundwater areas of the state | 2 | 8 | 8 | 0 | 1 |
| 4. Do you intend to be involved in the entire stakeholder process for groundwater? *n* = 55 | | | | | |
| Yes | 12 | 11 | 12 | 3 | 3 |
| No | 0 | 2 | 0 | 0 | 1 |
| Unsure | 0 | 2 | 7 | 1 | 1 |
| 5. In my opinion, the scope of the Groundwater Availability Assessment is appropriate. *n* = 55 | | | | | |
| Strongly agree | 3 | 6 | 2 | 2 | 1 |
| Agree | 5 | 8 | 13 | 1 | 3 |
| Disagree | 1 | 0 | 3 | 0 | 0 |
| Strongly disagree | 0 | 0 | 0 | 0 | 1 |
| Do not know | 3 | 1 | 1 | 1 | 0 |
| 6. I support additional resources for the Groundwater Availability Assessment and planning efforts. *n* = 54 | | | | | |
| Strongly agree | 4 | 5 | 5 | 2 | 3 |
| Agree | 8 | 9 | 8 | 2 | 2 |
| Disagree | 0 | 0 | 2 | 0 | 0 |
| Strongly disagree | 0 | 0 | 2 | 0 | 0 |
| Do not know | 0 | 1 | 1 | 0 | 0 |

| Question | Environmental, Conservation, or Nongovernmental Group | Government | Industry or Utility | Irrigated Uses | Other |
|---|---|---|---|---|---|
| 7.  Would you agree that you are now more informed on the Groundwater Availability Assessment? *n* = 54 | | | | | |
|     Strongly agree | 7 | 9 | 5 | 1 | 1 |
|     Agree | 4 | 6 | 12 | 2 | 2 |
|     Disagree | 0 | 0 | 1 | 0 | 0 |
|     Strongly disagree | 1 | 0 | 0 | 0 | 0 |
|     Do not know | 0 | 0 | 0 | 1 | 2 |
| 8.  Groundwater modeling provides useful information for groundwater users in the state. *n* = 54 | | | | | |
|     Strongly agree | 7 | 9 | 5 | 1 | 1 |
|     Agree | 4 | 6 | 12 | 2 | 2 |
|     Disagree | 0 | 0 | 1 | 0 | 0 |
|     Strongly disagree | 1 | 0 | 0 | 0 | 0 |
|     Do not know | 0 | 0 | 0 | 1 | 2 |
| 9.  Do you support statewide groundwater withdrawal permitting? *n* = 53 | | | | | |
|     Strongly support | 10 | 10 | 8 | 0 | 1 |
|     Do not support at all | 0 | 1 | 1 | 0 | 0 |
|     Permitting only where groundwater problems exist | 1 | 2 | 4 | 3 | 3 |
|     Not sure | 1 | 1 | 5 | 1 | 1 |
| 10. How important do you feel the Groundwater Availability Assessment is for water resources management? *n* = 54 | | | | | |
|     Very important | 12 | 10 | 15 | 4 | 3 |
|     Somewhat important | 0 | 3 | 3 | 0 | 2 |
|     Not important | 0 | 0 | 0 | 0 | 0 |
|     Do not know | 0 | 1 | 1 | 0 | 0 |

## APPENDIX 3:
## GROUNDWATER STAKEHOLDER RESPONSE BY REGISTERED/PERMITTED USER RESPONSE

| Question | Yes | No |
|---|---|---|
| 1. Which organization type do you represent? $n = 53$ | | |
| Environmental, conservation, or nongovernmental organization | 1 | 11 |
| Government | 2 | 13 |
| Industry or utility | 10 | 8 |
| Irrigated uses | 4 | 0 |
| Other | 0 | 4 |
| 2. My interests are mainly in _____. $n = 52$ | | |
| Groundwater availability | 4 | 3 |
| Groundwater quality | 0 | 0 |
| Groundwater contamination | 0 | 0 |
| Groundwater/surface water interactions | 1 | 2 |
| All of the above | 11 | 31 |
| 3. Which groundwater area do you represent? $n = 49$ | | |
| Inner coastal | 5 | 10 |
| Outer coastal | 7 | 8 |
| Outside coastal plains | 0 | 1 |
| All groundwater areas | 3 | 15 |
| 4. Do you intend to be involved in the entire stakeholder process for groundwater? $n = 53$ | | |
| Yes | 13 | 27 |
| No | 0 | 3 |
| Unsure | 4 | 6 |
| 5. In my opinion, the scope of the Groundwater Availability Assessment is appropriate. $n = 53$ | | |
| Strongly agree | 2 | 12 |
| Agree | 12 | 16 |
| Disagree | 2 | 2 |
| Strongly disagree | 0 | 1 |
| Do not know | 1 | 5 |

| Question | Yes | No |
|---|---|---|
| 6. I support additional resources for the Groundwater Availability Assessment and planning efforts. $n = 52$ | | |
| Strongly agree | 6 | 17 |
| Agree | 9 | 13 |
| Disagree | 0 | 2 |
| Strongly disagree | 0 | 1 |
| Do not know | 1 | 3 |
| 7. Would you agree that you are now more informed on the Groundwater Availability Assessment? $n = 52$ | | |
| Strongly agree | 6 | 12 |
| Agree | 7 | 21 |
| Disagree | 1 | 1 |
| Strongly disagree | 2 | 0 |
| Do not know | 0 | 2 |
| 8. Groundwater modeling provides useful information for groundwater users in the state. $n = 52$ | | |
| Strongly agree | 7 | 16 |
| Agree | 9 | 16 |
| Disagree | 0 | 1 |
| Strongly disagree | 0 | 1 |
| Do not know | 1 | 1 |
| 9. Do you support statewide groundwater withdrawal permitting? $n = 51$ | | |
| Strongly support | 7 | 21 |
| Do not support | 0 | 2 |
| Permitting only where groundwater problems exist | 5 | 7 |
| Not sure | 5 | 4 |
| 10. How important do you feel the Groundwater Availability Assessment is for water resources management? $n = 52$ | | |
| Very important | 16 | 27 |
| Somewhat important | 0 | 7 |
| Not important | 0 | 0 |
| Do not know | 1 | 1 |

Journal of South Carolina Water Resources, Volume 5, Issue 1, Pages 61–67, 2018

# Visualizing Relative Potential for Aquatic Ecosystem Toxicity Using the EPA Toxics Release Inventory and Life Cycle Assessment Methods

THEODORE LANGLOIS[1], MICHAEL CARBAJALES-DALE[1], AND ELIZABETH CARRAWAY[1]

AUTHORS: [1]Department of Environmental Engineering and Earth Sciences, Clemson University, 342 Computer Court, Anderson, SC 29625.

**Abstract.** The U.S. EPA Toxic Release Inventory has been available since 1987 as a record of industrial releases of toxic chemicals following the 1986 Emergency Planning and Community Right-to-Know Act. Combining this release data with estimates of relative toxicity of these chemicals to aquatic systems increases the value of the database by providing a common basis for comparison. The Tool for Reduction and Assessment of Chemicals and Other Environmental Impacts is a database of characterization factors to assess environmental impacts. It was used to develop relative ecotoxicity impacts and interpreted using Life Cycle Assessment concepts. The visualization software Tableau was used to generate representations of the preliminary results in this communication. The major potential sources of aquatic toxicity have been identified for South Carolina by industry type and by year over the period 1987–2016. The possibility of toxicity from releases of zinc compounds from power generation and pulp and paper mills far exceeds all other sources. Zinc compounds dominated the potential ecotoxicity over the full time period 1987–2016.

## INTRODUCTION

In response to the December 1984 industrial disaster at a Union Carbide plant in Bhopal, India, which released approximately 40 tonnes of methyl isocyanate ($CH_3NCO$) gas, and smaller scale industrial accidents in the United States, Congress passed the 1986 Emergency Planning and Community Right-to-Know Act or EPCRA (Broughton, 2005; Koehler & Spengler, 2006). The law addressed the potential for incidents that could affect human health in areas surrounding chemical or industrial plants. Section 313 of this statute charged the Environmental Protection Agency (EPA) with creating a list of facilities and their yearly releases of hazardous chemicals, the result being the Toxics Release Inventory (TRI). Since 1987, the EPA has maintained a list of toxic chemicals and thresholds that, if exceeded by a facility, must be reported. Over its 3 decades of existence, the most significant modification was the addition of more than 200 chemicals in 1994, bringing the list to more than 600 reportable chemicals and chemical categories. The resulting database offers the public itemized reports of masses of chemicals released into water, air, and soil by each facility, thus providing an annual summary of hazardous chemical releases by industrial activities. The TRI is currently available online for the years 1987 to 2016 (U.S.

EPA, 1987–2017). As legislation, EPCRA and the TRI initiated a new way of regulating industry; instead of an agency enforcing limits, the approach provides an information network that private citizens and interest groups can use to exert pressure on polluters until they reduce toxic waste to a level the public deems acceptable (Fung & O'Rourke, 2000). It is important to note that TRI does not track illegal releases; rather, it accounts for permitted releases associated with industrial processes. The program is generally agreed to be quite successful (Dahl, 1997; Ritter, 2015; Wolf, 1996). From 1988, the second year of the program, to 1995, the total amount of toxic chemicals released or transferred decreased by about 45% (U.S. EPA, 1987–2017).

Although it serves as a valuable tool for communities, the TRI does not reflect relative risks because toxicity information is absent within the database. Available data are presented as releases to water, air, and land by pound of chemical. Thus, using TRI data only, a user can compare releases of mercury compounds to lead compounds by mass, with no indication of which is potentially more harmful. To assess potential risk or damage to human and ecosystem health due to TRI-reported industrial releases, additional data, models, and more comprehensive analysis are needed.

To some degree, EPA has remedied this gap, through annual *TRI National Analysis* publications that analyze yearly release trends (U.S. EPA, 2018b) and through the creation of the Risk-Screening Environmental Indicators (RSEI), which is a model incorporating TRI data with measures of human exposure and toxicity (U.S. EPA, 2018a). The RSEI model assigns toxicity weights to chemicals based solely on human health effects. Additionally, the EPA in 2016 released a visualization tool to present TRI data and provide outreach for its Pollution Prevention (P2) program (Gaona & Kohn, 2016). The tool uses the visualization and mapping software QlikSense (qlik.com) to aid in visual analysis of large data sets and provide better tools to the public. Although powerful in its capabilities and accessibility to nonexperts, this specific tool, like the TRI itself, conveys only pounds of toxic waste managed.

For assessment of broader environmental impacts, EPA has developed the Tool for Reduction and Assessment of Chemicals and Other Environmental Impacts (TRACI; Bare 2011; Bare et al., 2002). TRACI provides factors for the estimation of chemical effects in several impact categories, for example, ozone depletion, global warming, acidification, eutrophication, and ecotoxicity. The current version of TRACI, Version 2.1, although available to the public on the EPA website, is primarily used by Life Cycle Impact Assessment practitioners and researchers (U.S. EPA, 2012). The use of TRACI requires input of data, such as TRI chemical releases; selection of options; and interpretation of impact characterization results. TRI data is not incorporated into this tool as it is in the RSEI model. TRACI also does not incorporate any visualization or mapping tools.

A few studies have combined TRI and TRACI to investigate broad applicability of the tools, as well as more specific LCA questions. No reports were found illustrating the combination of TRI, TRACI, and visualization software. Toffel and Marshall (2004) evaluated 13 weighting schemes for converting TRI data to potential environmental and human health effects and recommended EPA's products RSEI and TRACI. Lim et al. (2010) performed an in-depth analysis of 2007 TRI data coupled with human health and ecotoxicity potentials from TRACI. Their results showed that, in general, none of the chemicals identified as highest priority concerns using toxicity-based adjustments would be identified with TRI quantity-based data alone. Zhou and Schoenung (2009) illustrated the use of TRI data and an aggregation of impact assessment tools with a case study of the chemical manufacturing industry. Lam et al. (2011) identified pollution prevention options in the printed writing board manufacturing industry through analysis of TRI data, TRACI, and RSEI. Sengupta et al. (2015) examined ethanol and gasoline production processes using National Emissions Inventory data supplemented by TRI data and TRACI to estimate environmental and human health impacts.

Any combination of TRI and models such as TRACI generates numerical results and adds to the mass of environmentally related data available. To deal with large amounts of data, scientists, data analysts, and businesses are increasingly turning to visualization tools to provide data insights and inform decision making (Helbig et al., 2017; Palomino et al., 2017). These tools allow users to more easily extract important information from large datasets (Keim et al., 2008). It is a logical progression to use these tools to present data in an online and user-navigable format. This approach is consistent with the original mission of the TRI system, which is to provide the public with access to environmental data. The combination of visualization software with toxicity and environmental data can enhance the TRI program's availability and utility. Among several visualization software packages available, Tableau has been recognized as outstanding among commercial products (Nair et al., 2016). While the full Tableau product is a proprietary commercial product, Tableau Public (public.tableau.com) is a free product and online gallery that allows users to upload their visualizations and data sets for others to use or to connect to data files and create visualizations.

## PROJECT DESCRIPTION

This communication presents initial results in the development of an online visual data tool combining TRI data, TRACI ecotoxicity impact factors, LCA methodologies, and Tableau visualizations. The utility of this combination is illustrated for industrial toxic chemical releases to freshwater in South Carolina. LCA methodology was developed to help users understand relationships between the physical flow of chemicals and energy. Within the context of reports to TRI, it can be useful to combine LCA methods with a data management and visualization tool such as Tableau to generate innovative and useful data insights. With coincident freshwater resources and manufacturing industries, South Carolina represents an interesting case for the use of the combined tool.

LCA is generally reserved for evaluating the cradle-to-grave impacts of a product or system; however, it provides tools useful for analyzing environmental impacts on a local, statewide, and national scale (Zampori et al., 2016). LCA is composed of four phases: goal and scope definition, inventory analysis, impact assessment, and interpretation. In the inventory phase, elemental flows are tracked into and out of a product system. Raw materials, water, and energy may enter the boundaries of this system, while a final product and associated emissions exit. Although the TRI does not track products, it represents an inventory of chemical byproducts from manufacturing. In the impact assessment phase, an LCA practitioner uses inventory results to determine the types of impacts associated with releases to the environment.

These impacts belong to either midpoint or endpoint categories. Midpoint impacts are measurables that are directly influenced by chemical releases. For example, global warming potential is a midpoint category impacted by greenhouse gasses, whereas climate change is the endpoint impact related to global warming potential. Multiple midpoint impacts, such as aquatic ecotoxicity, acidity, and eutrophication, affect the ecosystem quality endpoint. Several models may be used to directly relate chemical releases into the environment with midpoint impacts. In this study, TRACI's Characterization Factors (CFs), which are based on chemical toxicity studies and environmental transport models, are used to assess potential environmental impacts in terms of a mass of a reference compound or relative units of toxicity (U.S. EPA, 2012).

## METHODS

In this analysis, direct-to-water releases from TRI are converted to ecotoxicity midpoint impact values using relevant CFs in the TRACI database. The final LCA phase, interpretation, is done through analysis and visualization using Tableau software (Version 2018.2, tableau.com). TRI and TRACI data were downloaded from the EPA website, compiled into Microsoft Access databases, and imported into Tableau for analysis (U.S. EPA, 1987–2017; U.S. EPA, 2012). The overall process is outlined as TRI data "inventory" × TRACI CFs = Tableau midpoint indicators.

Ecosystem toxicity, referred to in TRACI as ecotoxicity, is given by

$$CTU_e = W(kg) * CF\left(\frac{CTU_e}{kg}\right),$$

where $CTU_e$ is the comparative toxicity unit for ecotoxicity; W is the mass of chemical released according to the TRI database, measured in kilograms; and CF is the measure of ecotoxicity associated with each chemical in the TRACI database, measured in $CTU_e$/kg. $CTU_e$ values are proportional to estimates of potentially affected fractions of species, integrated over time and volume, per unit mass of a chemical emitted (USEtox, 2010). This calculation allows different chemicals to be compared in terms of their potential to harm species within an ecosystem. When multiplied together, using a Tableau data join and in-program calculation, the product is a comparative ecotoxicity value for each year and reporting location for each chemical or chemical class. The comparative nature of this ecotoxicity measure must be stressed; the $CTU_e$ is not a specific prediction of effects on species by a chemical; rather, it represents a method of relating expected ecotoxicity across a wide range of conditions and releases.

The TRACI database includes multiple CFs for different modes of release: to air (urban or rural), water (fresh or marine), and land (agricultural or natural soil). Some assumptions must be made to choose CFs and generate comparable results. First, we assume that all chemical releases to water in South Carolina in the TRI database were

made to freshwater and selected CFs for freshwater. Second, because TRI data groups certain metal compounds together and TRACI does not, a proxy compound must be chosen to represent a group of compounds. The RSEI methodology document (U.S. EPA, 2018) states that these compound categories are assumed to be metals in their most toxic form. Thus, the TRI category for "Copper Compounds" is associated with the TRACI chemical "Copper (II)."

## RESULTS

Figure 1 illustrates the dramatic difference in the chemicals that contribute most to TRI-reported releases to freshwater in South Carolina when assessed by mass and $CTU_e$. Results are shown for 2016. The data shown in Part A were adapted from the 2016 EPA National Analysis Results for South Carolina, whereas those presented in Part B are results generated by the tool developed in this project. Nitrate compounds clearly dominate by a wide margin in terms of mass of releases but do not appear when adjusted to reflect potential ecotoxicity effects.

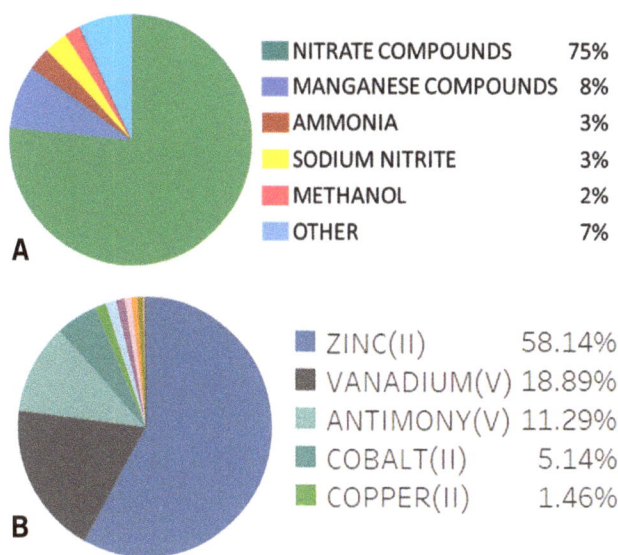

**Figure 1.** Top 5 releases to freshwater in South Carolina in 2016 by (A) mass as reported by the Toxics Release Inventory and (B) ecotoxicity.

For a broader perspective, Figure 2 presents the comparative toxicity (in millions of $CTU_e$) for total TRI-reported releases to freshwater in South Carolina between 1987 and 2016, grouped by industry sectors. A few industries and chemicals have dominated ecotoxicity to South Carolina's waterways over the past 30 years. It is clear that zinc compounds consistently present the largest ecosystem risk, especially from fossil fuel generation and the paper and pulp mill sectors. Four of the top 10 largest sources are related to paper or pulp manufacturing. Other significantly toxic releases include copper, vanadium, cobalt, and antimony compounds.

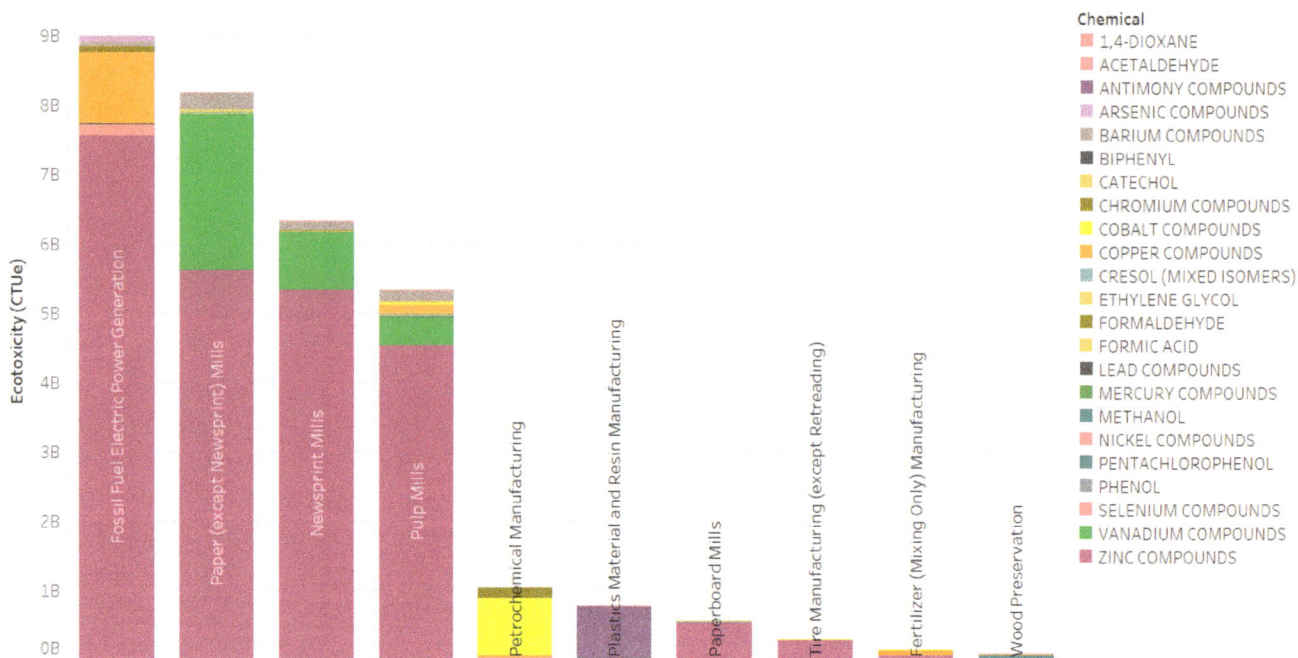

**Figure 2.** Top 10 industrial sectors releasing toxic chemicals to South Carolina waterways, 1987–2016. CTU = comparative toxicity unit.

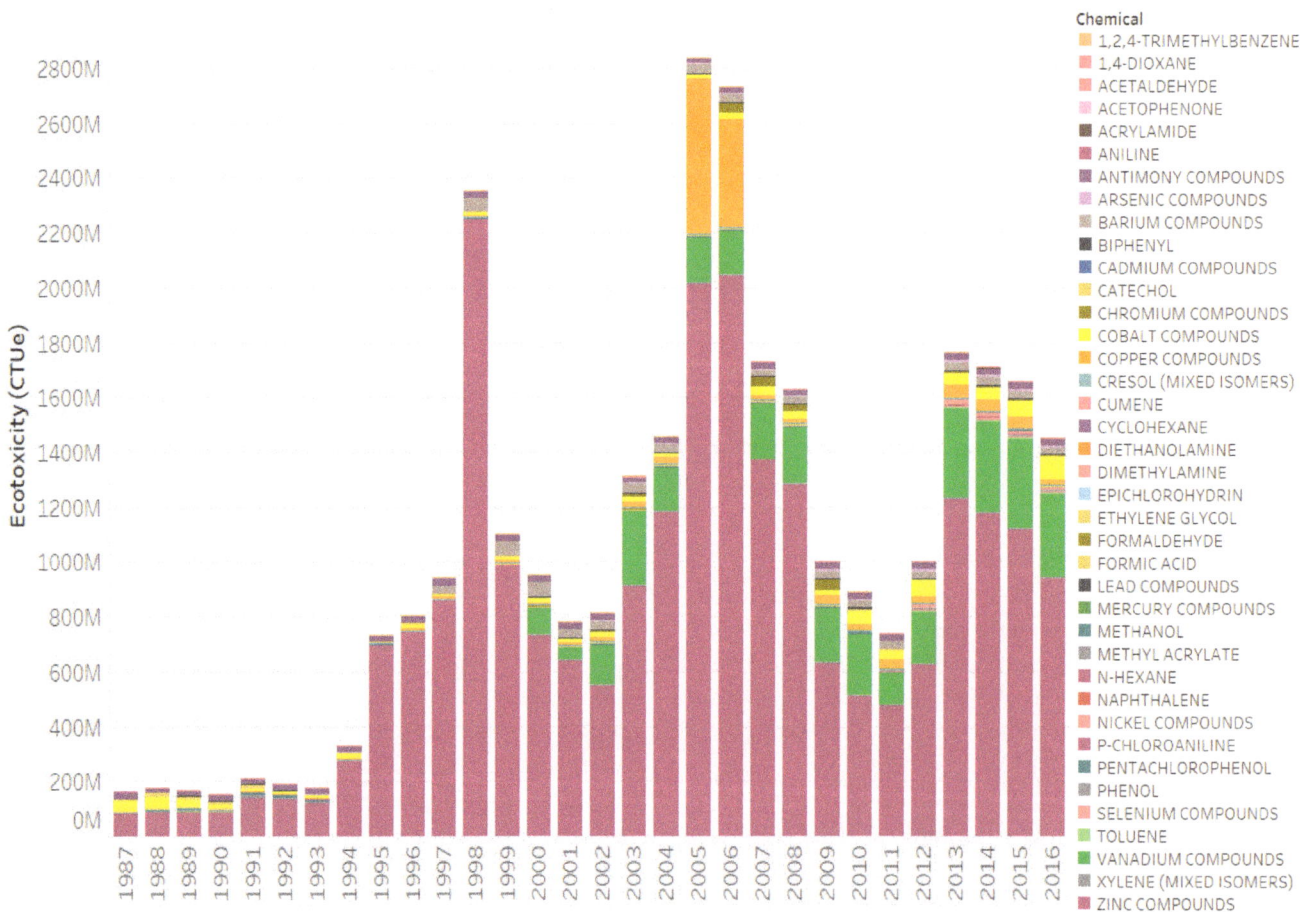

**Figure 3.** Annual variability of comparative ecotoxicity by chemical class. CTU = comparative toxicity unit.

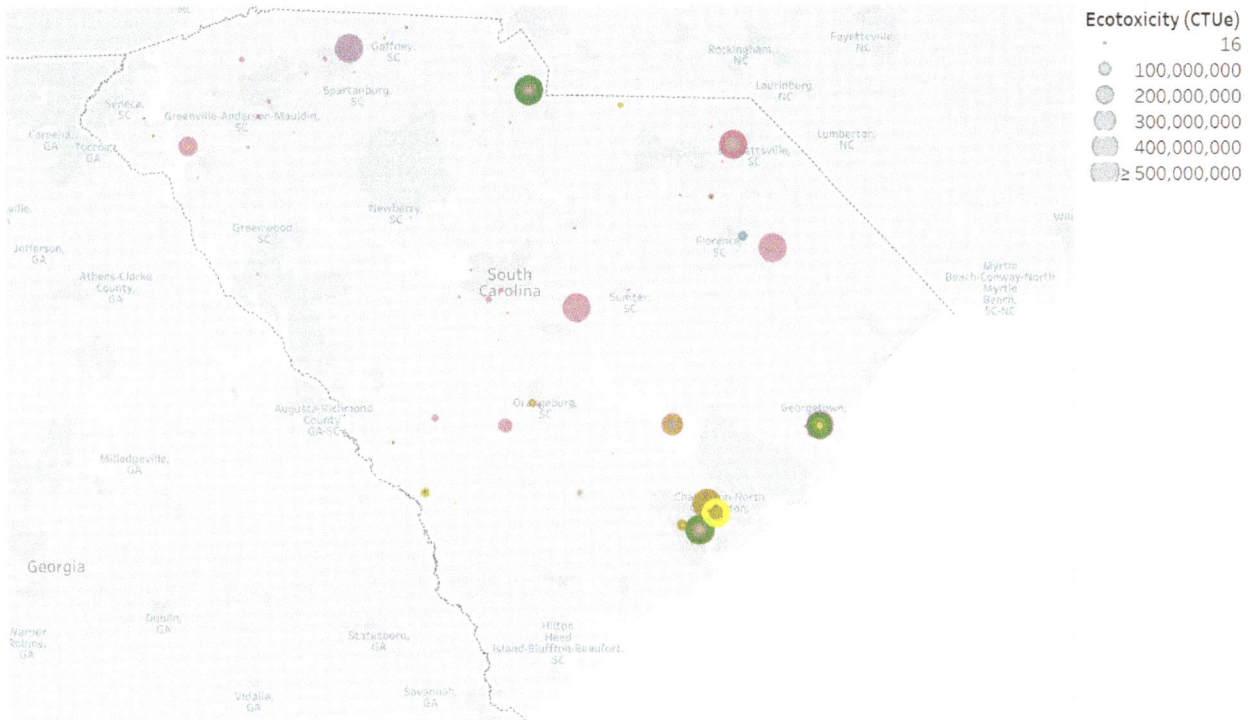

**Figure 4.** Comparative ecotoxicity of releases from South Carolina facilities, summed over 1987–2016. CTU = comparative toxicity unit.

Figure 3 shows the annual trend in ecotoxicity risks over the history of TRI data collection, with time on the $X$-axis and ecotoxicity measured in $CTU_e$ on the $Y$-axis. Vanadium compounds were added to the TRI list in 2000, adding to the overall yearly toxicity. Despite a general increase in production efficiency in the United States, the level of toxicity released to South Carolina water bodies increased in the late 1990s and experienced another increase in the mid-2000s, most likely due to an overall increase in manufacturing in the state. However, releases decreased sharply following the economic recession, which is reflected in this data (Koh et al. 2016).

Figure 4 maps locations of toxic chemical releases to South Carolina waters summed over 1987–2016. The distribution of TRI-reported releases aligns with major manufacturing areas in the state. There are concentrations in the Spartanburg–Greenville area, the Charlotte metro area, Georgetown, and Charleston. Many plants are near freshwater bodies used for recreation and drinking water supply.

Figure 5 presents the annual variability of comparative ecotoxicity from TRI-reported releases in South Carolina and the United States as a whole. Interestingly, the trends in ecotoxicity do not directly correlate between South Carolina

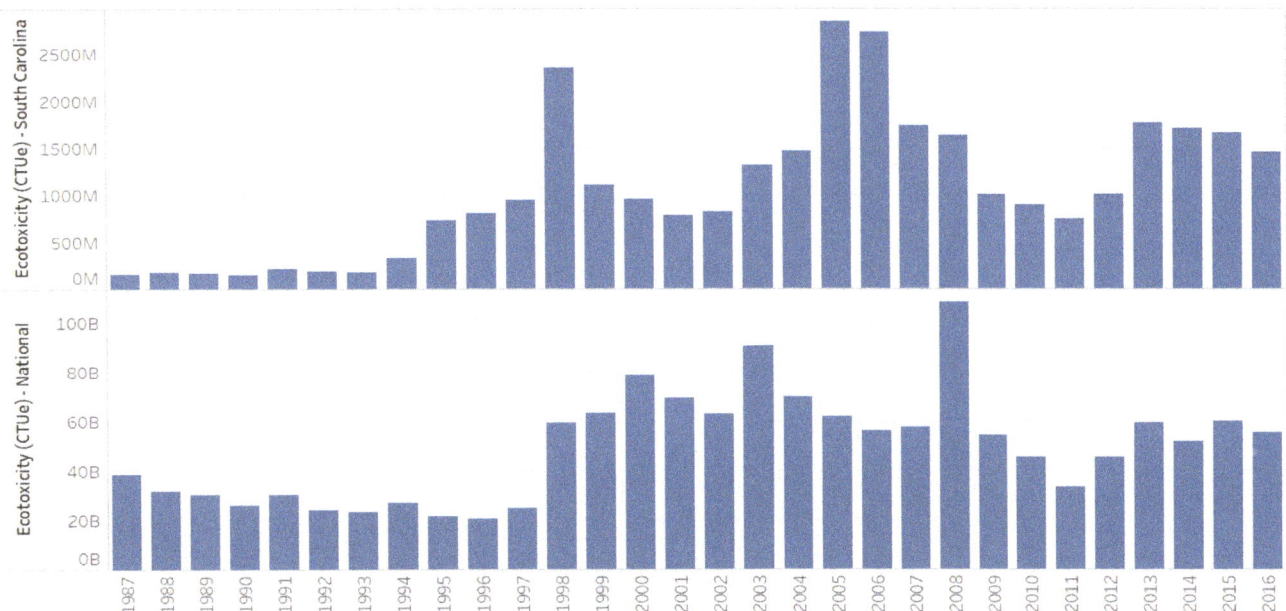

**Figure 5.** South Carolina and U.S. trends in comparative ecotoxicity, 1987–2016. CTU = comparative toxicity unit.

and the rest of the United States. While ecotoxicity in the early years of TRI declined in the United States, it remained relatively low and stable in South Carolina. The years 1998, 2005, and 2006 are significant in that they show sharp increases for South Carolina, while national trends were smoother or trending downward. These points could be further investigated to determine potential influences such as changes in reporting rules or activities of specific industries in the state. Finally, the state was consistent with the rest of the country with respect to the decline in operation and subsequent toxic releases after the financial crisis of 2008. Although release of hazardous materials can be tied to economic growth, especially for the manufacturing sector, it is, of course, not a desirable outcome. As South Carolina continues to grow its economy through industry, companies and private citizens should closely monitor environmental impacts of hazardous chemical release.

## DISCUSSION

The apparent variability in toxicity levels indicates potential problems with using TRI as a marker for gains or losses in environmental protection. First, the nature of the reporting mechanism places relatively little importance on accuracy. It is estimated that in its first year, 10,000 out of 30,000 facilities required to comply with the program failed to do so, and in any given year, only 3% of facilities are investigated by EPA (Wolf, 1996). Second, the sitting EPA administration has the power to add and remove chemicals on a year-by-year basis. This means that the chemical list from 1987 differs significantly from the 2016 list. Third, chemicals can change reporting categories. In one year, a chemical release or method of treatment may be listed in a different category from the next. This creates a phantom or paper reduction, which appears as a decrease in trends but does not in fact correspond to a physical reduction (Natan & Miller, 1998). Despite reporting errors, changing categories, and adding or removing chemicals, the TRI database is a valuable source for tracking industrial chemical releases. The illustrations of using LCA and visualization techniques for freshwater releases in South Carolina show the importance of including toxicity factors when assessing potential impacts to ecosystems. In particular, this analysis predicts that chemicals containing zinc exert more harm than those containing nitrate.

## ACKNOWLEDGEMENTS

The authors wish to thank the Pollution Prevention Program of U.S. EPA Region 4 for support of P2 activities related to this work.

## LITERATURE CITED

Bare, JC. 2011. TRACI 2.0: the tool for the reduction and assessment of chemical and other environmental impacts 2.0. Clean Technol Environ Policy. 13(5):687–696.

Bare JC, Norris GA, Pennington DW, McKone T. 2002. TRACI: the tool for the reduction and assessment of chemical and other environmental impacts. J Ind Ecol. 6(3–4):49–78.

Broughton E. 2005. The Bhopal disaster and its aftermath: a review. Environ Health. 4(1):6.

Dahl R. 1997. Now that you know. Environ Health Perspect. 105(1):38–42.

Fung A, O'Rourke D. 2000. Reinventing environmental regulation from the grassroots up: explaining and expanding the success of the Toxics Release Inventory. Environ. Manage. 25(2):115–127.

Gaona S, Kohn J. 2016. Using the visualization software Qlik for TRI data presentation and P2 outreach. Washington (DC): U.S. EPA. https://www.epa.gov/sites/production/files/2016-11/documents/gaona_qlik_for_tri_data_presentation_and_p2_outreach.pdf.

Helbig C, Dransch D, Bottinger M, Devey C, Haas A, Hlawitschka M, Kuenzer C, Rink K, Schafer-Neth C, Scheuermann G, et al. 2017. Challenges and strategies for the visual exploration of complex environmental data. Int J Digit Earth. 10(10):1070–1076.

Keim D, Andrienko G, Fekete J-D, Görg C, Kohlhammer J. 2008. Visual analytics: definition, process and challenges. In: Kerren A, Stasko JT, Fekete J-D, North C, editors. Information visualization: human-centered issues and perspectives. New York (NY): Springer. p. 154–175.

Koehler D, Spengler J. 2006. The toxic release inventory: fact or fiction? A case study of the primary aluminum industry. Environ. Manage. 85:296–307.

Koh SCL, Ibn-Mohammed T, Acquaye A, Feng K, Reaney IM, Hubacek K, Fujii H, Khatab K. 2016. Drivers of U.S. toxicological footprints trajectory 1998–2013. Sci. Rep. 6:39514.

Lam CW, Lim SR, Schoenung JM. 2011. Environmental and risk screening for prioritizing pollution prevention opportunities in the U.S. printed wiring board manufacturing industry. J. Hazard. Mater. 189(1–2):315–322.

Lim SR, Lam CW, Schoenung JM. 2010. Quantity-based and toxicity-based evaluation of the U.S. Toxics Release Inventory. J. Hazard. Mater. 178(1–3):49–56.

Nair LR, Shetty SD, Shetty SD. 2016. Interactive visual analytics on big data: Tableau vs. D3.JS. J. e-Learning Knowl. Soc. 12(4):139–150.

Natan TE, Miller CG. 1998. Are toxic release inventory reductions real? Environ. Sci. Technol. 32(15):368A–374A.

Palomino J, Muellerklein OC, Kelly M. 2017. A review of the emergent ecosystem of collaborative geospatial tools for addressing environmental challenges. Comput Environ Urban Syst. 65:79–92.

Ritter SK. 2015. EPA analysis suggests green success. Chem. Eng. News. 93(5):32–33.

Sengupta D, Hawkins TR, Smith RL. 2015. Using national inventories for estimating environmental impacts of products from industrial sectors: a case study of ethanol and gasoline. Int J Life Cycle Assess. 20(5);597–607.

Toffel MW, Marshall JD. 2004. Improving environmental performance assessment: a comparative analysis of weighting methods used to evaluate chemical release inventories. J Ind Ecol. 8(1–2):143–172.

U.S. Environmental Protection Agency (EPA). 1987–2017. TRI basic data files: calendar years 1987–2017. www.epa.gov/toxics-release-inventory-tri-program/tri-basic-data-files-calendar-years-1987-2017.

U.S. Environmental Protection Agency (EPA). 2012. Tool for the reduction and assessment of chemical and other environmental impacts (TRACI): TRACI version 2.1: user's guide. https://nepis.epa.gov/Adobe/PDF/P100HN53.pdf.

U.S. Environmental Protection Agency (EPA). 2018a. EPA's risk-screening environmental indicators (RSEI) methodology. RSEI Version 2.3.6. www.epa.gov/rsei.

U.S. Environmental Protection Agency (EPA). 2018b. Toxics Release Inventory (TRI) national analysis. www.epa.gov/trinationalanalysis.

USEtox. 2010. USEtox user manual. www.usetox.org/support/tutorials-manuals.

Wolf, SM. 1996. Fear and loathing about the public right to know: the surprising success of the Emergency Planning and Community Right-to-Know Act. J. Land Use Environ. Law. 11:217–324.

Zampori L, Saouter E, Castellani V, Schau E, Cristobal J, Sala S. 2016. Guide for interpreting life cycle assessment result. (JRC technical reports, EUR 28266 EN).

Zhou X, Schoenung JM. 2009. Combining U.S.-based prioritization tools to improve screening level accountability for environmental impact: the case of the chemical manufacturing industry. J. Hazard. Mater. 172(1):423–431.

Journal of South Carolina Water Resources, Volume 5, Issue 1, Pages 69–73, 2018

# An Online Tool for Estimating Evapotranspiration and Irrigation Requirements of Crops in South Carolina

JOSÉ O. PAYERO[1*]

AUTHOR: [1]Irrigation Specialist, Clemson University Edisto Research and Education Center, Blackville, SC, 29817.
*jpayero@clemson.edu

**Abstract.** In recent years, there has been an increased interest in South Carolina regarding the amount of water used by different consumers, especially agricultural producers. This interest has sparked conversations among different stakeholders, including the media, policy makers, producers, scientists, and the general public, regarding the current state and future of water resources in the state. Central to these discussions, from the agricultural sector perspective, is the question of how much water producers really need to grow crops. The objective of this study was, therefore, to develop an online tool to use local South Carolina historic weather data to estimate daily and seasonal crop evapotranspiration and irrigation requirements for different crops. The overall goal was for the new tool to assist farmers and other stakeholders to better plan irrigation water allocations and management. Therefore, an interactive online tool called ETcCalc was created to address this objective. ETcCalc, which is freely available online (http://sccropwater.com), was developed using historic weather data; therefore, it is suitable as an irrigation planning tool rather than a real-time irrigation scheduling tool.

## INTRODUCTION

In South Carolina, water use for irrigation is mostly unregulated compared with other states. Currently, only some areas of the state are classified as capacity use areas, where water users with the capacity to withdraw over 3 million gallons in any given month are required to obtain a permit. In recent years, there has been considerable controversy in South Carolina regarding the unregulated use of water for agriculture. This has motivated legislators to consider imposing additional regulations on water use in the state. For example, in 2017, the South Carolina Department of Health and Environmental Control conducted a series of public hearings aimed at expanding the classification of capacity use areas, which had mainly impacted coastal counties, to cover some inland counties in the state. Recent events suggest that new legislations and regulations on agricultural water use are to be expected in years to come.

One of the critical questions that will be asked when developing new water regulations will be how much water a farmer really needs to grow a specific crop at a specific location in South Carolina. The answer to this question is complex, because the amount of water used by a crop, known as crop evapotranspiration (ETc), is heavily dependent on the local weather conditions, which vary by time of day, by day of the year, and from year to year. Furthermore, the seasonal irrigation requirements, in addition to the water used by the crop, also depend on other factors, such as soil type, soil water content at the time of planting, efficiency of the irrigation system and, especially, effective rainfall during the crop growing season. The amount of effective rainfall and ETc during the crop growing season are the two most important components influencing the irrigation requirements of a crop. The amount of rainfall can easily be measured with rain gauges, but directly measuring ETc is difficult and expensive.

The traditional method of measuring ETc is by planting the crop inside a weighing lysimeter (Figure 1) and measuring the changes in lysimeter mass during a given time interval (Fisher, 2012; Payero & Irmak, 2008; Schneider et. al., 1998). Changes in mass during relative short periods of time (hourly or daily) are assumed to be due to changes in water content of the soil inside the lysimeter box, which allow calculation of ETc. Lysimeters, however, are expensive to build, difficult to maintain, and fixed to a specific field.

In recent decades, micrometeorological methods, such as the eddy covariance (EC) method (Burba & Anderson, 2007), have become popular among researchers because they offer accuracy and portability. EC systems are similar to a weather station, which can be installed in the middle of a

Figure 1. Lysimeter installation (top), and cotton crop planted on lysimeter (bottom).

field and can be moved to a different field as needed (Figure 2). The EC system, for example, can measure all of the components of the 1-dimensional energy balance equation,

$$Rn - G = LE + H, \qquad (1)$$

where Rn is net radiation, G is soil heat flux, H is sensible heat flux, and LE is latent heat flux (all in units of W m$^{-2}$). ETc is derived by converting LE to units of water depth (inches or millimeters). EC systems, however, are very expensive (around $50,000 each), which severely limits their use to measure ETc outside a small number of research applications.

Although there is a long history of research on actual ETc measurements for different crops under different environments around the world (Tolk et al., 1998; Evett et al., 2009; Payero & Irmak, 2013), there is always a need for more measurements to keep pace with the development of new crop varieties and with the introduction of crops to new environments. Actual ETc measurements are also needed to calibrate and fine tune methods to estimate ETc from weather variables. However, it would be impossible to have actual measurements for every crop and every location. In the absence of actual local measurements of ETc, the next best thing is to estimate ETc from local weather

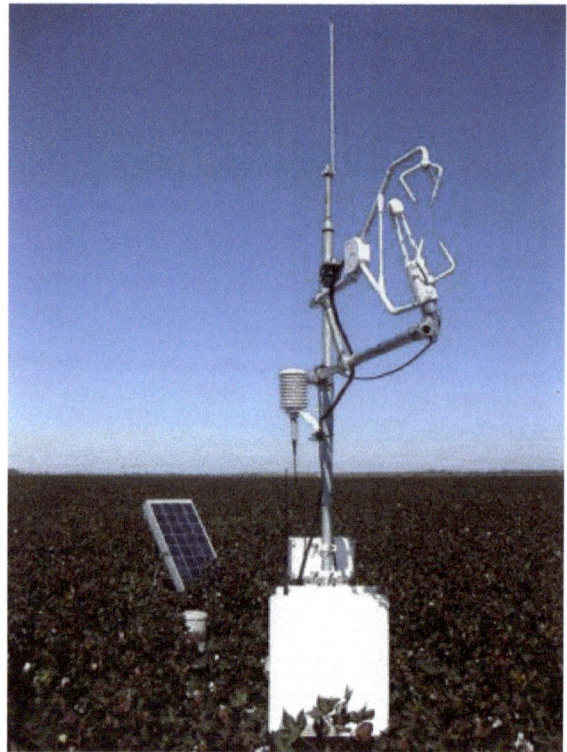

Figure 2. Eddy covariance system.

measurements. The objective of this study was, therefore, to develop an online tool to incorporate local South Carolina historic weather data to estimate daily and seasonal ETc and irrigation requirements for different crops. The overall goal was for the new tool to assist farmers and other stakeholders to better plan irrigation water allocation and management.

## METHODOLOGY

An interactive online tool called ETcCalc was created to address this objective and is freely available at http://sccropwater.com (Figure 3). Users can create a project and add up to 5 scenarios. Each scenario consists of a combination of a crop, location, planting date, and soil type.

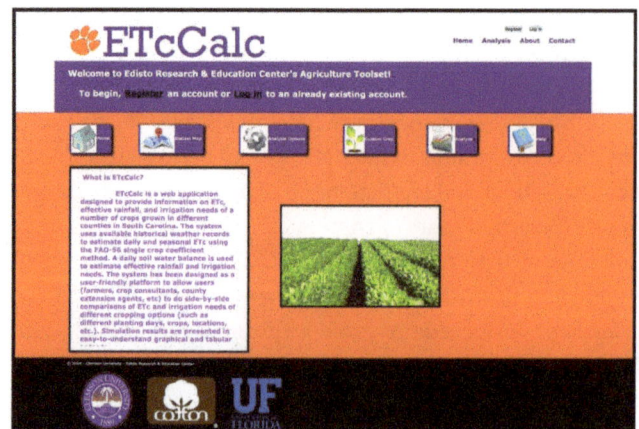

Figure 3. ETcCalc online tool.

For example, one scenario could include corn planted on April 10 in Aiken County in a silt loam soil, whereas another scenario could include soybean planted on June 12 in Anderson County in a sandy loam soil. The location can be chosen from a map (Figure 4) showing the weather stations in South Carolina that measure all the weather variables needed to calculate ETo using the Penman–Montheith method (solar radiation, air temperature, relative humidity, and wind speed). In 2014, we found that 696 weather stations were operating in South Carolina as part of the public weather station network, but only those stations shown in Figure 4 were equipped to measure all the variables needed to calculate ETo.

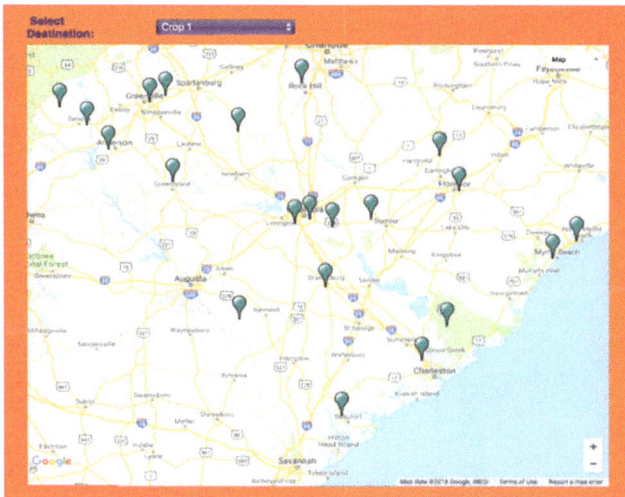

**Figure 4.** Location of weather stations in South Carolina included in ETcCalc.

After selecting the weather station from the map, the user can complete inputs for each scenario (Figure 5).

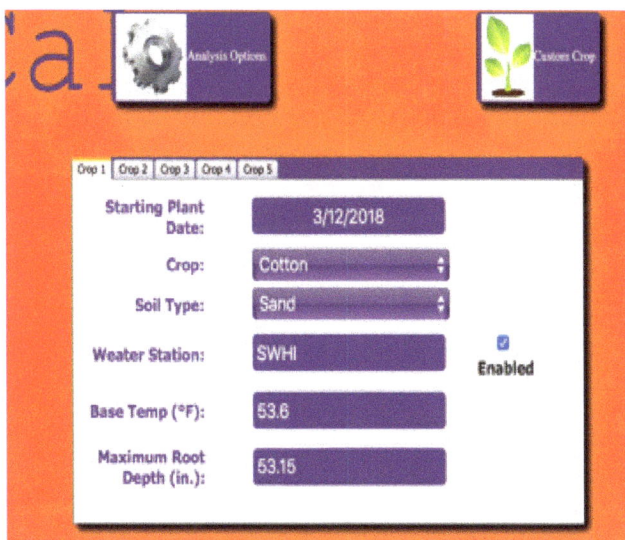

**Figure 5.** Inputs for each analysis scenario.

ETcCalc calculates daily ETc values for a crop with no water stress, using the Food and Agriculture Organization of

The United Nations' Paper 56 (FAO-56), "Guidelines for Computing Crop Water Requirements" (Allen et al., 1998) single crop coefficient procedure (Wright, 1982) as

$$ETc = Kc \times ETo, \qquad (2)$$

where ETc is crop evapotranspiration (inch per day), Kc is the crop coefficient, and ETo is grass-reference evapotranspiration (inch per day). Historic daily ETo values for each of the weather stations were obtained from the North Carolina Climate Retrieval and Observations Network of the Southeast Database (CRONOS; http://climate.ncsu.edu/cronos/), provided by the North Carolina Climate Office. These historic ETo values were then stored in a local database for easy access by ETcCalc. CRONOS calculates daily ETo values from weather data using the FAO-56 Penman–Montheith method (Allen et al., 1998). ETcCalc provides default Kc and length of growth stage (LGS) values for 19 of the main crops in South Carolina, which have been taken from FAO-56. Actual measurements of daily water use, Kc, and LGS values of local crops are severely lacking in South Carolina. However, this subject is currently under investigation, and default Kc and LGS values in ETcCalc will be updated as more local data become available.

In addition to the default Kc and LGS values, ETcCalc allows users to create new crops or new crop varieties by providing adequate Kc and LGS values (Figure 6) for the initial, development, midseason, and late-season development stages, as defined by FAO-56 (Table 1).

**Table 1.** Definition of crop grow stages according to the Food and Agriculture Organization.

| Crop Stage | Stage Definition |
|---|---|
| Initial | Planting to 10% ground cover |
| Development | 10% Ground cover to effective full cover |
| Midseason | Effective full cover to start of maturity |
| Late Season | Start of maturity to harvest or full senescence |

After the scenarios are specified in ETcCalc, an analysis can be performed. The resulting outputs would then show a side-by-side comparison of results from the different scenarios. ETcCalc calculates daily ETc values for every day in the specified historic weather record. Daily rainfall and ETc values are then used to conduct a daily soil water balance to estimate monthly and seasonal ETc, rainfall, effective rainfall (rain that is stored in the soil profile and is available to the crop), and rainfall deficit (ETc – effective rainfall). As a bonus, the tool also calculates daily growing degree days. ETcCalc presents results in both graphical and tabular formats.

## RESULTS

In the following section, the outputs of the tool are illustrated by creating several scenarios comparing the impact of changing planting date for cotton.

An analysis with 3 scenarios was conducted to illustrate some of the outputs of ETcCalc. The scenarios included cotton planted on 3 different planting dates (April 15, May 15, and June 15) in Orangeburg, South Carolina. Figures 7 and 8 show the daily ETc and cumulative ETc for the 3 cotton planting dates. They show that planting date can have a big impact on both daily and seasonal ETc.

Figures 9 and 10 show the monthly and seasonal summaries of ETc, rain, effective rain, and rain deficit for each of the 3 planting dates. The seasonal summary (Figure 10) indicates that cotton planted earlier in the compared scenarios would have more ETc and less effective rainfall and would, therefore, require more irrigation.

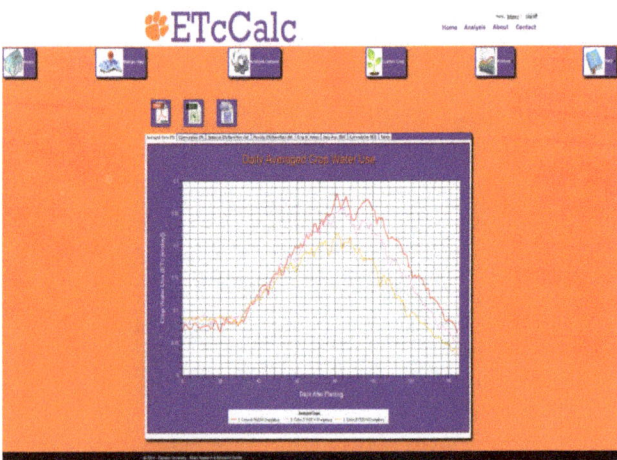

Figure 6. Inputs for each analysis scenario.

Figure 7. Daily crop evapotranspiration for cotton planted on 3 planting dates in Orangeburg, South Carolina.

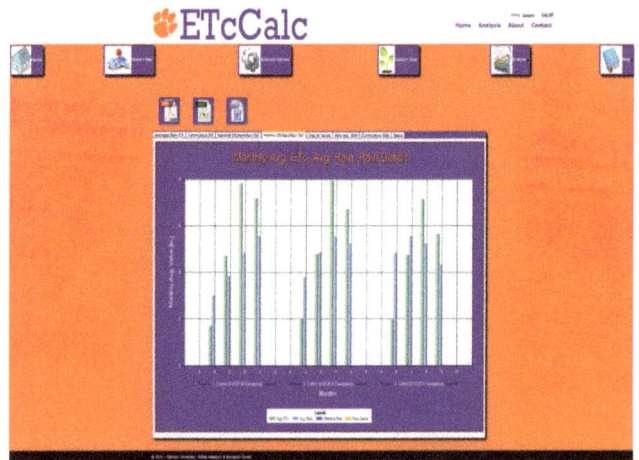

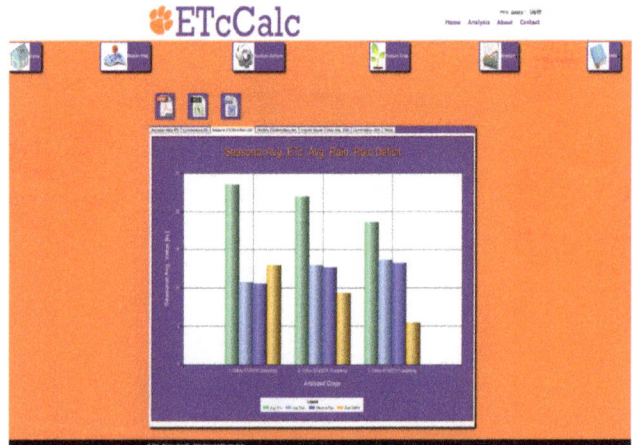

Figure 8. Daily cumulative crop evapotranspiration for cotton planted on 3 planting dates in Orangeburg, South Carolina.

Figure 9. Monthly summary of crop evapotranspiration (ETc), rain, effective rain, and rain deficit for each of the 3 scenarios.

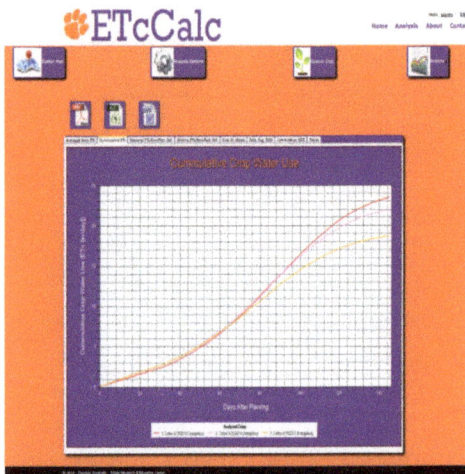

Figure 10. Seasonal summary of crop evapotranspiration (ETc; green), rain (light blue), effective rain (dark blue), and rain deficit (yellow) for each of the 3 scenarios.

## CONCLUSIONS

ETcCalc is a tool that facilitates calculation of crop ETc and irrigation requirement for crops and allows side-by-side comparisons of different cropping scenarios. The ETcCalc tool was initially developed and made available online in 2014; therefore, the historic weather dataset after that year is not included. Currently, a new online tool is under development that will expand on the capabilities of ETcCal and will link directly to the CRONOS database for automatic download of the latest weather and ETo data. A couple of online tools for real-time irrigation scheduling, rather than for irrigation planning, are also currently under development. The new irrigation scheduling tools will use real-time weather data, rather than the static historic dataset used by ETcCalc. One of the tools is being designed to use real-time weather data from CRONOS, and the other will use data that come from Weather Underground (www.wunderground.com).

## ACKNOWLEDGEMENTS

This is Technical Contribution Number 6640 of the Clemson University Experiment Station. This material is based upon work supported by U.S. Department of Agriculture, National Institute of Food and Agriculture, under Project Number SC-1700540. The author acknowledges Christopher Harvey for developing the computer code and the partial funding provided by the South Carolina Cotton Board.

## LITERATURE CITED

Allen RG, Pereira LS, Raes D, Smith M. 1998. Crop evapotranspiration: guidelines for computing crop water requirements (FAO irrigation and drainage paper 56). Rome (Italy): Food and Agriculture Organization of the United Nations. http://www.fao.org/docrep/X0490E/X0490E00.htm.

Burba GG, Anderson D. 2007. Introduction to the eddy covariance method: general guidelines, and conventional workflow. Lincoln (NE): Li-Cor Biosciences.

Evett SR, Mazahrih NT, Jitan MA, Sawalha MH, Colaizzi PD, Ayars JE. 2009. A weighing lysimeter for crop water use determination in the Jordan Valley, Jordan. Trans. ASABE, 52(1):155–169.

Fisher DK. 2012. Simple weighing lysimeters for measuring evapotranspiration and developing crop coefficients. Int. J. Agric. Biol. Eng. 5(3): 35–43.

Payero JO, Irmak S. 2008. Construction, installation, and performance of two repacked weighing lysimeters. Irrigation Sci. 26(2):191–202.

Payero JO, Irmak S. 2013. Daily energy fluxes, evapotranspiration and crop coefficient of soybean. Agric. Water Manag. 129:31–43.

Schneider AD, Howell TA, Moustafa ATA, Evett SR, Abou-Zeid W. 1998. A simplified weighing lysimeter for monolithic or reconstructed soils. Appl. Eng. Agric. 14(3):267–274.

Tolk JA, Howell TA, Evett SR. 1998. Evapotranspiration and yield of corn grown on three High Plains soils. Agron J. 90:447–454.

Wright JL. 1982. New evapotranspiration crop coefficients. J. Irrigation Drain. Div. ASCE. 108:57–74.

www.ingramcontent.com/pod-product-compliance
Lightning Source LLC
Chambersburg PA
CBHW061225270326
41927CB00025B/3500